T0320764

Fundamentals of Lightning

Written by one of the authors of the acclaimed reference volume *Lightning: Physics and Effects* (Cambridge, 2003), this new book provides a focused set of topics suitable for advanced undergraduate or graduate courses on lightning. It presents the current state of the art in lightning science, including areas such as lightning modeling, calculation of lightning electromagnetic fields, electromagnetic methods of lightning location, and lightning damaging effects and protective techniques.

Pedagogical features designed to facilitate class learning include end-of-chapter summaries, further reading suggestions, questions and problems, and a glossary explaining key lightning and atmospheric electricity terms. A selection of appendices is provided at the end of the book, which includes detailed derivations of exact equations for computing electric and magnetic fields produced by lightning.

Designed for a single-semester course on lightning and its effects, and written in a style accessible to technical non-experts, this book will also be a useful, up-to-date reference for scientists, engineers, and practitioners who have to deal with lightning in their work.

Vladimir A. Rakov is a professor in the Department of Electrical and Computer Engineering at the University of Florida and Co-Director of the International Center for Lightning Research and Testing (ICLRT). He is the author or coauthor of 3 other books and over 250 journal papers on various aspects of lightning. Professor Rakov has been elevated to the grade of Fellow by the Institute of Electrical and Electronics Engineers (IEEE), the American Meteorological Society (AMS), the American Geophysical Union (AGU), and the Institution of Engineering and Technology (IET). He received the Karl Berger Award from the Scientific Committee of the International Conference on Lightning Protection (ICLP) for distinguished achievements in lightning research, developing new fields in theory and practice, modeling and measurements.

Fundamentals of Lightning

Written in one of the author's the acclaimed reference volumes *Lightning: Physics and Effects* (Cambridge, 2003), this new book provides a focused and compact source of information ...

Fundamentals of Lightning

VLADIMIR A. RAKOV

Department of Electrical and Computer Engineering
University of Florida

CAMBRIDGE
UNIVERSITY PRESS

Shaftesbury Road, Cambridge CB2 8EA, United Kingdom

One Liberty Plaza, 20th Floor, New York, NY 10006, USA

477 Williamstown Road, Port Melbourne, VIC 3207, Australia

314–321, 3rd Floor, Plot 3, Splendor Forum, Jasola District Centre, New Delhi – 110025, India

103 Penang Road, #05–06/07, Visioncrest Commercial, Singapore 238467

Cambridge University Press is part of Cambridge University Press & Assessment,
a department of the University of Cambridge.

We share the University's mission to contribute to society through the pursuit of
education, learning and research at the highest international levels of excellence.

www.cambridge.org
Information on this title: www.cambridge.org/9781107072237

First published 2016

A catalogue record for this publication is available from the British Library

Library of Congress Cataloging-in-Publication data
Rakov, Vladimir A., 1955– author.
Fundamentals of lightning / Vladimir A. Rakov, Department of Electrical and
Computer Engineering, University of Florida.
Cambridge, United Kingdom ; New York, NY : Cambridge University Press, 2016. | ©2016 | Includes
bibliographical references and index.
LCCN 2016008231 | ISBN 9781107072237 (hardback ; alk. paper) | ISBN 1107072239
(hardback ; alk. paper)
LCSH: Lightning.
LCC QC966 .R34 2016 | DDC 551.56/32–dc23
LC record available at http://lccn.loc.gov/2016008231

ISBN   978-1-107-07223-7   Hardback

Additional resources for this publication at www.cambridge.org/rakov

To my wife Lucy,
our son Sergei and his wife Yulia,
and their children Alexander, Elizabeth, and Victor

Contents

Preface

This book, *Fundamentals of Lightning* (FOL), can be viewed as a condensed and updated version of *Lightning: Physics and Effects* (LPE), a 687-page monograph that was co-authored by Professor M. A. Uman and me and published by Cambridge University Press in 2003. As such, FOL draws heavily on LPE, although the new book contains a considerable amount of new material, particularly in Chapters 5, 7, 8, and 9 and in Appendices 3–5 and 7–9. Also, most of the material has been reorganized and brought up to date.

LPE (still in print) covers essentially all aspects of lightning, including lightning physics, lightning protection, and the interaction of lightning with a variety of objects and systems, as well as with the environment. It has become an important reference (over 1,000 citations, according to Google Scholar) on lightning and its effects for professionals and students alike. LPE has also been used as the textbook for a one-semester course, "Lightning," at the University of Florida (UF) and for other similar university courses worldwide. UF currently offers both graduate and senior undergraduate versions of the course, with typical total enrollment being 30 to 40. In using LPE as the textbook for over 10 years, I found it to be less than optimal in that it contains significantly more material than can be taught in a one-semester Lightning course, while lacking some material (for example, measurement of lightning electric and magnetic fields) covered by the course. Additionally, the LPE material is structured differently from the course syllabus. The content and the structure of the new, condensed book are harmonized with those of the UF Lightning course and, I hope, this will be appreciated by students. FOL includes a number of pedagogical enhancements: questions and problems at the end of each chapter, a glossary explaining basic lightning and atmospheric electricity terms, end-of-chapter summaries (points to remember), and further reading suggestions. Appendix 3 contains detailed derivations of exact equations for computing electric and magnetic fields produced by lightning, as per the recommendation of one of the reviewers of the book proposal. The prerequisite for the Lightning course at UF, which is part of the Electromagnetics curriculum, is an undergraduate course in electromagnetics, but an undergraduate general physics course that covers electromagnetics in moderate detail would be acceptable too.

Besides students, the book will be a valuable resource for scientists, engineers and other practitioners who have to deal with lightning in their work. Most of the book (notable exceptions might be Chapter 5 and Appendix 3) is written in a style that is accessible to the technical non-expert and should be useful for anyone interested in understanding lightning and its effects. In fact, Appendices 5–8 are written based largely on my general public lectures and interaction with mass media.

I would like to thank colleagues who read various parts of the manuscript, or write-ups that ended up in the manuscript, and provided useful comments and suggestions, including, in alphabetical order, K. L. Cummins, J. R. Dwyer, R. L. Holle, E. P. Krider, A. Nag, M. Rubinstein, R. Thottappillil, and M. D. Tran. Thanks also go to those who provided original figures included in the book: R. B. Anderson, Y. Baba, J. D. Hill, A. Yu. Kostinsky and E. A. Mareev, G. Maslowski, A. Nag, R. E. Orville, D. D. Sentman and W. A. Lyons, S. Sumi, and M. D. Tran. The cover image was obtained by M. D. Tran at the Lightning Observatory in Gainesville (LOG), Florida. Help with drafts of some chapters/appendices of the book and/ or figures was given by Y. Li, S. Mallick, W. Shi, N. N. Slyunyaev, V. B. Somu, M. D. Tran, and, in particular, P. Sun.

It has taken about two years to write *Fundamentals of Lightning*. During that time, I have benefited from my interaction with my current and former graduate students, S. Mallick, Dr. C. T. Mata, Dr. A. Nag, V. B. Somu, M. D. Tran, and Y. Zhu, and with my current and past Visiting Scholars, Prof. Y. Baba, Prof. Y. Chen, Dr. Y. Li, Prof. W. Lu, Prof. G. Maslowski, Dr. W. Shi, Dr. (soon to be) P. Sun, and Prof. D. Wang.

Miss Zoë Pruce, Assistant Editor of the Earth and Environmental Science division of Cambridge University Press, provided an outstanding level of support, particularly at the final stages of preparation of the manuscript.

1 Types of lightning discharges and lightning terminology

Lightning can be defined as a transient, high-current (typically tens of kiloamperes) electric discharge in air whose length is measured in kilometers. As for any discharge in air, the lightning channel is composed of ionized gas – that is, of plasma – whose peak temperature is typically 30,000 K—about five times higher than the temperature of the surface of the Sun. Lightning was present on Earth long before human life evolved, and it may even have played a crucial role in the evolution of life on our planet. The global lightning flash rate is some tens per second to 100 per second. Each year, some 25 million cloud-to-ground lightning discharges occur in the United States, and this number is expected to increase by about 50 percent due to global warming over the twenty-first century (Romps et al., 2014). Lightning initiates many forest fires, and over 30 percent of all electric power line failures are lightning related. Each commercial aircraft is struck by lightning on average once a year. A lightning strike to an unprotected object or system can be catastrophic.

1.1 Overview

The lightning discharge in its entirety, whether it strikes the ground or not, is usually termed a "lightning flash" or just a "flash." A lightning discharge that involves an object on the ground or in the atmosphere is referred to as a "lightning strike." A commonly used non-technical term for a lightning discharge is a "lightning bolt." About three quarters of lightning discharges do not involve the ground. They include intracloud, intercloud, and cloud-to-air discharges and are collectively referred to as cloud flashes (see Fig. 1.1) and sometimes as ICs. Lightning discharges between cloud and earth are termed cloud-to-ground (or just ground) discharges and sometimes referred to as CGs. The latter constitute about 25 percent of global lightning activity. About 90 percent or more of global cloud-to-ground lightning is accounted for by negative downward lightning, in which negative charge is effectively transported to the ground, and the initial process begins in the cloud and develops in the downward direction. The term "effectively" is used to indicate that individual charges are not transported all the way from the cloud to the ground during the lightning processes; rather, the flow of electrons (the primary charge carriers) in one part of the lightning channel results in the flow of other electrons in other parts of the channel. Other types of cloud-to-ground lightning include positive downward, negative upward, and positive upward discharges (see Fig. 1.2). Downward flashes exhibit downward branching, while upward flashes are branched upward. Upward lightning discharges (types (b) and (d) in Fig. 1.2) are thought to occur only from tall objects (higher than 100 m or so) or from

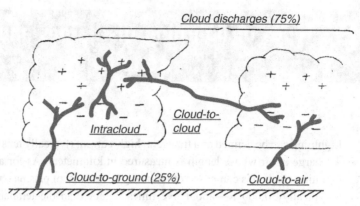

Types of lightning discharges from cumulonimbus

Fig. 1.1 General classification of lightning discharges from thunderstorm clouds. Cloud discharges constitute 75 percent and cloud-to-ground discharges 25 percent of global lightning activity.

objects of moderate height located on mountain tops. There are also bipolar lightning discharges sequentially transferring both positive and negative charges during the same flash. Bipolar lightning discharges are usually initiated from tall objects (that is, are of the upward type). Downward bipolar lightning discharges do occur, but appear to be rare.

Cloud flashes are most likely to begin near the upper and lower boundaries of the main negative charge region, and in the former case often bridge the main negative and main positive charge regions in the cloud (see chapter 3). Other scenarios are possible. There is a special type of cloud lightning that is thought to be the most intense natural producer of HF–VHF (3–300 MHz) radiation on Earth. It is referred to as compact intracloud discharge (CID). CIDs received their name due to their relatively small (hundreds of meters) spatial extent. They tend to occur at high altitudes (mostly above 10 km), appear to be associated with strong convection (however, even the strongest convection does not always produce CIDs), and tend to produce less light than other types of lightning discharges. Additional information on CIDs is given in Appendix 4.

Lightning occurrence is not limited to the Earth's atmosphere. There exists convincing evidence for lightning or lightning-like discharges on Jupiter and Saturn. Currents in Jovian lightning are expected to be one to two orders of magnitude larger than in Earth lightning. A review of extraterrestrial lightning is given by Rakov and Uman (2003, Ch. 16).

1.2 Downward negative lightning

We first introduce, referring to Figs. 1.3a and b, the basic elements of the negative downward lightning discharge, termed "component strokes" or just "strokes." Each flash typically contains three to five strokes, the observed range being one to 26.

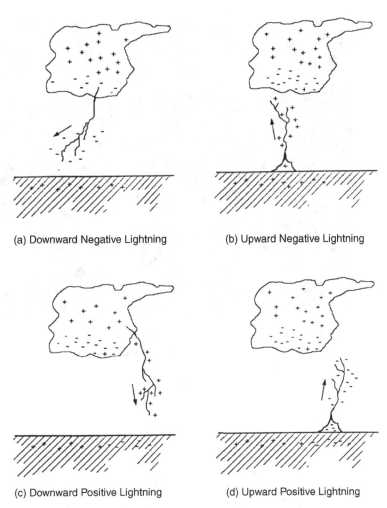

(a) Downward Negative Lightning   (b) Upward Negative Lightning

(c) Downward Positive Lightning   (d) Upward Positive Lightning

Fig. 1.2   Four types of lightning effectively lowering cloud charge to the ground. Only the initial leader is shown for each type. For each lightning-type name given below the sketch, direction (downward or upward) indicates the direction of propagation of the initial leader and polarity (negative or positive) refers to the polarity of the cloud charge effectively lowered to the ground. In (a) and (c), the polarity of charge lowered to the ground is the same as the leader polarity, while in (b) and (d) those polarities are opposite. Not shown in this figure are upward (object-initiated) and downward bipolar lightning flashes. © Vladimir A. Rakov and Martin A. Uman 2003, published by Cambridge University Press, reprinted with permission.

Roughly half of all lightning discharges to earth strike the ground at more than one point, with the spatial separation between the channel terminations being up to many kilometers. Then we will introduce, referring to Figs. 1.4a and b, the two major lightning processes comprising a stroke, the "leader" and the "return stroke," which occur as a sequence with the leader preceding the return stroke. We will also briefly review lightning parameters, with more details being found in Rakov and Uman (2003) and references therein.

Fig. 1.3   Lightning flash which appears to have at least seven (perhaps as many as ten) separate ground strike points: (a) still-camera photograph, (b) moving-camera photograph. Some of the strike points are associated with the same stroke, having separate branches touching the ground, while others are associated with different strokes taking different paths to the ground. Adapted from Hendry (1993).

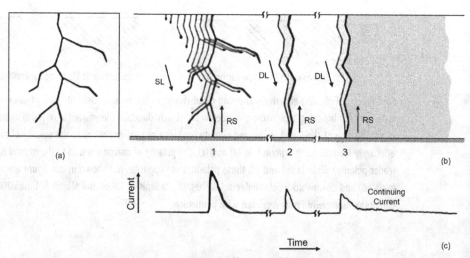

Fig. 1.4   Schematic diagram showing the luminosity of a three-stroke downward negative flash and the corresponding current at the channel base: (a) still-camera image, (b) streak-camera image, and (c) channel-base current. © Vladimir A. Rakov and Martin A. Uman 2003, published by Cambridge University Press, reprinted with permission.

Two photographs of a negative cloud-to-ground discharge are shown in Figs. 1.3a and b. The image in Fig. 1.3a was obtained using a stationary camera, while the image in Fig. 1.3b was captured with a separate camera that was moved horizontally during the time of the flash. As a result, the latter image is time-resolved, showing several distinct luminous channels between the cloud and the ground separated by dark gaps. The distinct channels are associated with individual strokes, and the time intervals corresponding to the dark gaps are typically of the order of tens of milliseconds. These dark time intervals between strokes explain why lightning often appears to the human eye to "flicker." In Fig. 1.3b, time advances from right to left, so that the first stroke is on the far right. The first two strokes are branched, and the downward direction of branches indicates that this is a downward lightning flash.

Now we consider sketches of still and time-resolved images of the three-stroke lightning flash shown in Figs. 1.4a and b, respectively. A sketch of the corresponding current at the channel base is shown in Fig. 1.4c. In Fig. 1.4b, time advances from left to right, and the timescale is not continuous. Each of the three strokes in Fig. 1.4b, represented by its luminosity as a function of height above ground and time, is composed of a downward-moving process, termed a "leader," and an upward-moving process, termed a "return stroke (RS)." Transition from the leader to the return stroke is referred to as the "attachment process," which is not shown in Fig. 1.4b. The leader creates a conducting path between the cloud charge source region and the ground and distributes negative charge from the cloud source region along this path, and the return stroke traverses that path moving from the ground toward the cloud charge source region and neutralizes the negative leader charge. Thus, both leader and return stroke processes serve to effectively transport negative charge from the cloud to the ground. As seen in Fig. 1.4b, the leader initiating the first return stroke differs from the leaders initiating the two subsequent strokes (all strokes other than first are termed "subsequent strokes"). In particular, the first-stroke leader appears optically to be an intermittent process – hence the term "stepped leader (SL)" – while the tip of a subsequent-stroke leader appears to move continuously. The continuously moving subsequent-stroke leader tip appears on streak photographs as a downward-moving "dart," hence the term "dart leader (DL)." The apparent difference between the two types of leaders is related to the fact that the stepped leader develops in virgin air, while the dart leader follows the "pre-conditioned" path of the preceding stroke or strokes. Sometimes, a subsequent leader exhibits stepping while propagating along a previously formed channel; it is referred to as "dart-stepped leader." There are also so-called "chaotic" subsequent-stroke leaders.

The electric potential difference between a downward-moving stepped-leader tip and the ground is probably some tens of megavolts, comparable to or a considerable fraction of that between the cloud charge source and the ground. The magnitude of the potential difference between two points, one at the cloud charge source and the other on the ground, is the line integral of electric field intensity between those points. The upper and lower limits for the potential difference between the lower boundary of the main negative charge region and the ground can be estimated by multiplying, respectively, the typical observed electric field in the cloud, 10^5 V m^{-1}, and the expected electric field at the ground under a thundercloud immediately prior to the initiation of lightning, 10^4 V m^{-1}, by the height of the lower boundary of the negative charge region above ground. The resultant range is 50–500 MV, if the height is assumed to be 5 km.

When the descending stepped leader attaches to the ground, the first return stroke begins. The first-return-stroke current measured at the ground rises to an initial peak of about 30 kA in some microseconds and decays to half-peak value in some tens of microseconds. The return stroke effectively lowers to the ground the several coulombs of charge originally deposited on the stepped-leader channel, including all the branches. Once the bottom of the dart leader channel is connected to the ground, the second (or any subsequent) return-stroke wave is launched upward, which again serves to neutralize the leader charge. The subsequent return-stroke current at the ground typically rises to a peak value of 10–15 kA in less than a microsecond and decays to half-peak value in a few tens of microseconds.

The high-current return-stroke wave rapidly heats the channel to a peak temperature near or above 30,000 K and creates a channel pressure of 10 atm (1 megapascal) or more, resulting in channel expansion, intense optical radiation, and an outward propagating shock wave that eventually becomes the thunder (sound wave) we hear at a distance. Each cloud-to-ground lightning flash involves an energy of roughly 10^9 to 10^{10} J (1–10 gigajoules). Lightning energy is approximately equal to the energy required to operate five 100 W light bulbs continuously for one month. Note that not all the lightning energy is available at the strike point; only 10^{-2}–10^{-3} of the total energy, since most of the energy is expended in producing thunder, hot air, light, and radio waves.

The impulsive component of the current in a return stroke is often followed by a continuing current, which has a magnitude of tens to hundreds of amperes and a duration of up to hundreds of milliseconds. Continuing currents with a duration in excess of 40 ms are traditionally termed "long continuing currents." Between 30 and 50 percent of all negative cloud-to-ground flashes contain long continuing currents. Current pulses superimposed on continuing currents, as well as the corresponding enhancements in luminosity of the lightning channel, are referred to as M-components.

1.3 Downward positive lightning

Positive lightning discharges are relatively rare (about 10 percent of global cloud-to-ground lightning activity), but there are five situations that appear to be conducive to the more frequent occurrence of positive lightning. These situations include: (1) the dissipating stage of an individual thunderstorm; (2) winter (cold-season) thunderstorms; (3) trailing stratiform regions of mesoscale convective systems; (4) some severe storms; and (5) thunderclouds formed over forest fires or contaminated by smoke. Positive flashes are usually composed of a single stroke, in contrast with negative flashes, about 80 percent of which contain two or more strokes, with three to five being typical. Multiple-stroke positive flashes do occur but are relatively rare. Positive lightning is typically more energetic and potentially more destructive than negative lightning.

The gross charge structure of a "normal" thundercloud is often viewed as a vertical tripole consisting of three charge regions with the main positive at the top, main negative in the middle, and an additional (typically smaller) positive below the main negative (see chapter 3). Such a charge structure appears to be not conducive to the production of positive

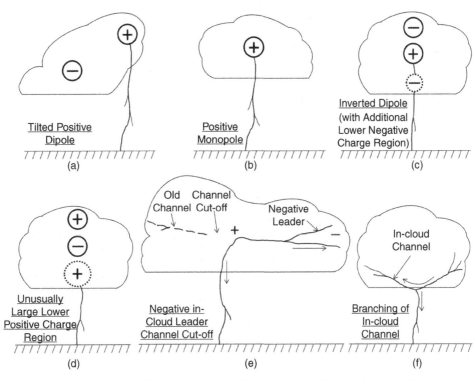

Fig. 1.5 Conceptual cloud charge configurations and scenarios leading to production of downward positive lightning. Adapted from Nag and Rakov (2012).

cloud-to-ground lightning. Figure 1.5 illustrates six conceptual cloud charge configurations and scenarios that were observed or hypothesized to give rise to positive lightning. For four of them, tilted positive dipole, positive monopole, inverted dipole, and unusually large lower positive charge region (Figs. 1.5a–1.5d, respectively), the primary source of charge is a charged cloud region, while for the other two, negative in-cloud leader channel cutoff and branching of in-cloud channel (Figs. 1.5e and 1.5f, respectively), the primary source of charge is an in-cloud lightning channel formed prior to the positive discharge to the ground.

1.4 Artificially initiated lightning

Lightning can be initiated artificially (triggered) by launching a small rocket trailing a thin grounded or ungrounded wire toward a charged cloud overhead. In the former case, the triggered lightning is referred to as "classical" and in the latter case as "altitude." The sequence of processes (except for the transition from leader to return stroke stage that is referred to as the "attachment process") in classical triggered lightning is schematically

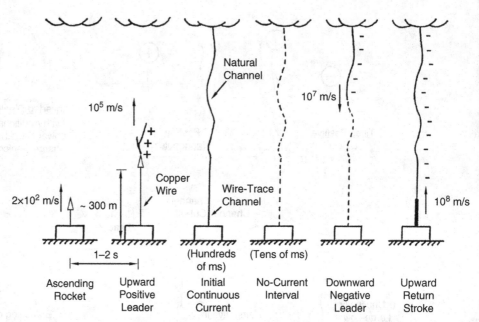

Fig. 1.6 Sequence of events (except for the attachment process) in classical triggered lightning. The upward positive leader and initial continuous current constitute the initial stage of a classical triggered flash. Adapted from Rakov et al. (1998).

shown in Fig. 1.6. When the rocket, ascending at about 150–200 m s^{-1}, is about 200–300 m high, the field enhancement near the rocket tip launches a positively charged leader that propagates upward toward the cloud. This upward positive leader vaporizes the trailing wire, bridges the gap between the cloud and the ground, and establishes an initial continuous current with a duration of some hundreds of milliseconds that transports negative charge from the cloud charge source region to the triggering facility. After the cessation of the initial continuous current, one or more downward dart-leader/upward return-stroke sequences may traverse the same path to the triggering facility. The dart leaders and the following return strokes in triggered lightning are similar to dart-leader/return-stroke sequences in natural lightning, although the initial processes in natural downward and triggered lightning are distinctly different. To date, well over 1,000 lightning flashes have been triggered by researchers in different countries using the rocket-and-wire technique, with over 450 of them at the International Center for Lightning Research and Testing (ICLRT) at Camp Blanding, Florida. The ICLRT was established in 1993 and, since 2004, also includes the Lightning Observatory in Gainesville (LOG), located 45 km from the Camp Blanding facility. Photographs of two classical rocket-and-wire triggered-lightning flashes from Camp Blanding are shown in Fig. 1.7.

The results of triggered-lightning experiments have provided considerable insight into natural lightning processes that would not have been possible from studies of natural lightning due to its random occurrence in space and time. As an example, Fig. 1.8 shows a photograph of surface arcing during a triggered-lightning flash from

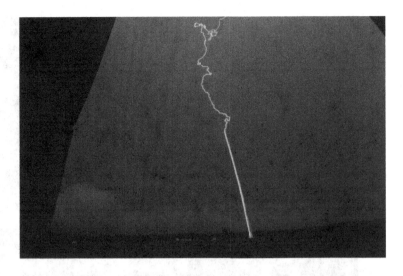

Fig. 1.7 Photographs of lightning flashes triggered using the rocket-and-wire technique at Camp Blanding, Florida. Top – a distant view of a strike to the test runway; bottom – a close-up view of a strike to the test power system. © Vladimir A. Rakov and Martin A. Uman 2003, published by Cambridge University Press, reprinted with permission.

experiments at Fort McClellan, Alabama. The soil was red clay and a 0.3 or 1.3 m steel vertical rod was used for grounding the rocket launcher. The surface arcing appears to be random in direction and often leaves little if any evidence on the ground. Even within the same flash, individual strokes can produce arcs developing in different directions. In one case, it was possible to estimate the current carried by one arc branch which contacted the instrumentation. That current was approximately 1 kA, or 5 percent of the total current peak in that stroke. The observed horizontal

Fig. 1.8 Photograph of surface arcing associated with the second stroke (current peak of 30 kA) of flash 9312 triggered at Fort McClellan, Alabama. Lightning channel is outside the field of view. One of the surface arcs approached the right edge of the photograph, a distance of 10 m from the rocket launcher. Adapted from Fisher et al. (1994).

extent of surface arcs was up to 20 m, which was the limit of the photographic coverage during the Fort McClellan experiments. These results suggest that the uniform ionization of soil, usually postulated in studies of the behavior of grounding electrodes subjected to lightning surges, may not be an adequate assumption.

1.5 Upward lightning

The phenomenology of upward negative lightning, illustrated in Fig. 1.9 by the sketches of still and time-resolved photographic records and of the corresponding current record, is similar to that of negative lightning triggered using the classical rocket-and-wire technique (Section 1.4). In the latter case, the thin triggering wire plays the role of the grounded object, one that is rapidly erected and then replaced by the plasma channel of the upward leader. Downward leader/upward return stroke sequences in upward lightning (Fig. 1.9), and their counterparts in rocket-triggered lightning (Fig. 1.6), are similar to subsequent strokes in natural downward lightning (Fig. 1.4). For this reason, leader/return stroke sequences in upward (object-initiated) lightning and in rocket-triggered lightning are sometimes referred to as "subsequent strokes." Interestingly, only 20 to 50 percent of object-initiated flashes contain subsequent strokes, while over 70 percent of rocket-and-wire triggered flashes in Florida do so. Upward flashes can be negative, positive, or bipolar. A photograph of an upward flash (note upward branching) initiated from a 70 m tower on a 640 m mountain is shown in Fig. 1.10.

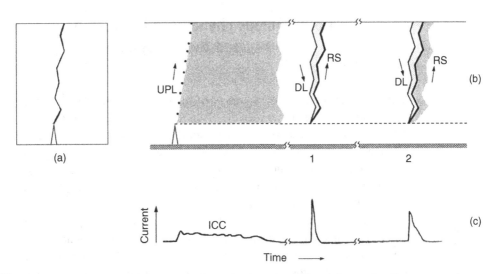

Fig. 1.9 Schematic diagram showing the luminosity of an upward negative flash and the corresponding current at the channel base. (a) Still-camera image, (b) streak-camera image, (c) current record. The flash is composed of an upward positive leader (UPL) followed by an initial continuous current (ICC) and two downward dart leader (DL)/upward return stroke (RS) sequences. The UPL and ICC constitute the initial stage of an upward negative flash. © Vladimir A. Rakov and Martin A. Uman 2003, published by Cambridge University Press, reprinted with permission.

Fig. 1.10 Photograph of an upward flash initiated from a 70 m tower on Monte San Salvatore, Switzerland. Courtesy of R. E. Orville.

1.6 Luminous phenomena between cloud tops and the ionosphere

Lightning discharges cause a variety of transient, relatively-low-luminosity optical phenomena in the clear air between the cloud tops (at altitudes of near 20 km or less) and the lower ionosphere (near 60–90 km depending on the time of day). Six general types of such phenomena, which are collectively referred to as transient luminous events (TLEs), have been observed: red sprites, halos, blue starters, blue jets, gigantic jets, and elves (see Fig. 1.11). They represent a mechanism for energy transfer from lightning and the thundercloud to the regions of the atmosphere between the cloud tops and the lower ionosphere. Sprites have tens-of-kilometers vertical extent and complex spatial structure. From spatially and temporally averaged observations, they have peak optical intensities generally between 0.1 and 10 MR (megarayleigh; optical intensity of aurora can reach 1 MR) and overall durations from about a millisecond to many tens of milliseconds. Sprites are difficult to see, even with a dark-adapted eye. Halos are brief descending glows with lateral extent 40–70 km, which accompany or precede (but not always) more structured sprites. Blue starters and blue jets propagate upward from the cloud tops at speeds near 10^5 m/s. The

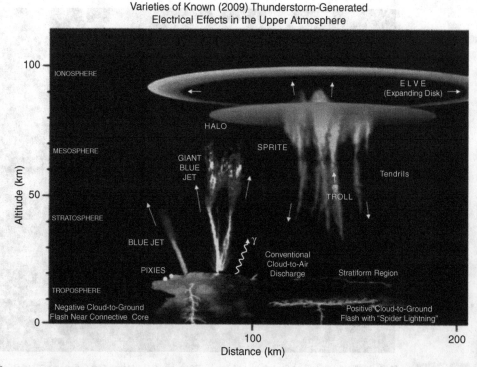

Fig. 1.11   Illustration of transient luminous events occurring between cloud tops and the lower ionosphere. Courtesy of D. D. Sentman and W. A. Lyons.

lower, brightest part of blue starters and blue jets apparently can have optical intensities above 10 MR, while their tops are considerably less bright. Starters extend less than 10 km above cloud tops – generally only a few kilometers – while blue jets have vertical extents of 20 km or so. Blue starter and blue jet durations are up to a few hundreds of milliseconds. Occasionally, jet-like events propagate all the way to the ionosphere, in which case they are referred to as gigantic jets. Elves expand outward across the lower ionosphere in less than a millisecond to a maximum horizontal extent of 200–700 km and are reported to have optical intensities of roughly 1–10 MR.

1.7 High-energy radiation from lightning and thunderstorms

Recent observations of hard X-rays and gamma-rays in thunderstorms (other than the enhancements of gamma-ray background due to precipitation of radon by rain) fall into three categories: surges in the gamma-ray background lasting seconds to minutes; bursts of X-rays associated with all kinds of natural and triggered lightning leaders, typically occurring within less than 1 ms of the return stroke; and Terrestrial Gamma-ray Flashes or TGFs (typically less than 1 ms in duration). The latter are usually observed from space, but on a few occasions have been seen on the ground or from an aircraft. At present, the only viable mechanism for producing energetic radiation involves runaway electrons, which occur when the energy gained by the free electrons between collisions (as they are accelerated by a high electric field) exceeds the energy that is lost by collisions with air molecules. X-rays and gamma-rays are produced via what is called bremsstrahlung (braking radiation) that is emitted when a free electron is deflected in the electric field of a nucleus or, to a lesser extent, in the field of an atomic electron.

The energy spectrum of observed gamma-ray glows is consistent with the relativistic runaway electron avalanche (RREA) mechanism (also referred to as the relativistic runaway breakdown theory), which requires energetic seed electrons (of the order of 0.1–1 MeV) produced by cosmic rays and sufficiently high electric fields (calculated to be of the order of 100 kV/m at an altitude of 6 km) extending over a sufficient distance (of the order of a kilometer). For all types of negative lightning leaders, the energy of individual X-ray photons was estimated to be in the 30 to 250 keV range (the upper limit is twice the energy of a chest X-ray), although occasionally photons in the MeV range were observed. It is likely that X-ray emissions from cloud-to-ground lightning leaders are associated with the so-called cold runaway (also known as thermal runaway) breakdown, in which very strong electric fields (>30 MV/m) cause the high-energy tail of the bulk-free electron population to grow, allowing some electrons to run away to high energies. Such very high fields may be present at streamer heads or leader tips. It does not appear that the runaway breakdown is a necessary feature of lightning leaders. TGFs are associated with thunderstorms and individual lightning flashes, with accompanying electromagnetic signals (sferics), mostly suggesting intracloud flashes effectively transporting negative charge upward, including some CIDs, as the type of

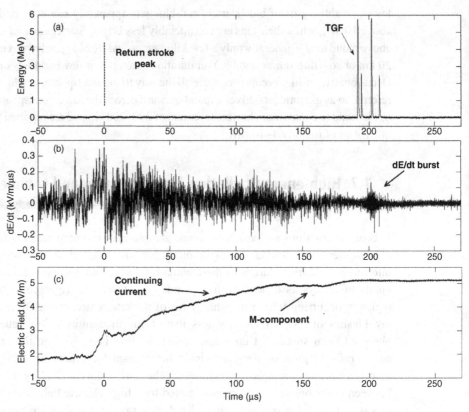

Fig. 1.12 (a) TGF observed at the LOG on June 13, 2014. It had a duration of 16 μs (six X-ray/gamma-ray pulses, two of which can only be seen on an expanded vertical scale), and began 191 μs after the return-stroke electric field peak. (b) dE/dt and (c) low-gain electric field records acquired at the LOG (7.5 km from the lightning channel). The NLDN-reported return-stroke peak current was 224 kA. Adapted from Tran et al. (2015).

parent lightning. As of today, it is not clear which lightning process is responsible for the production of TGFs. One possibility is the cold runaway breakdown during the stepping process of a negative in-cloud leader. On the other hand, according to Dwyer and Cummer (2013), TGFs could be produced in the absence of ordinary lightning via runaway breakdown processes (RREAs – relativistic runaway electron avalanches – in the large-scale homogeneous electric field inside the cloud) alone. Since the latter processes emit little visible light, the phenomenon was referred to as "dark lightning."

Presented in Fig. 1.12 is an example of TGF observed at ground level. It was recorded at the Lightning Observatory in Gainesville (LOG), Florida, and was associated with a single-stroke negative cloud-to-ground discharge. A total of six photons were detected, two of which exceeded the upper measurement limit of 5.7 MeV. The peak current reported by the US National Lightning Detection Network (NLDN) for the return stroke preceding the TGF was as high as 224 kA.

1.8 The global electric circuit

The electrical conductivity of the air at sea level is about 10^{-14} S m^{-1}, and it increases rapidly with altitude. It is usually assumed that the atmosphere above a height of 60 km or so, under quasi-static conditions, becomes conductive enough to consider it equipotential. This region of atmosphere is sometimes referred to as the electrosphere. The potential of the electrosphere is positive with respect to the Earth and its magnitude is about 300 kV, with most of the voltage drop taking place below 20 km where the electric field is relatively large. The overall situation is often visualized as a lossy spherical capacitor (e.g. Uman, 1974), the outer and inner shells of which are the electrosphere and Earth's surface, respectively. According to this model, the Earth's surface is negatively charged with the total charge magnitude being roughly 5×10^5 C, while an equal amount of positive charge is distributed throughout the atmosphere. Because the atmosphere between the capacitor "shells" is weakly conducting, there is a fair-weather leakage current of the order of 1 kA (2 pA m^{-2}; 1 pA = 10^{-12} A) between the shells that would neutralize the charge on the Earth and in the atmosphere on a timescale of roughly 10 minutes (depending on the amount of pollution) if there were no charging mechanism to replenish the neutralized charge. Since the capacitor is observed to remain charged, there must be a mechanism or mechanisms acting to resupply that charge. Wilson (1920) suggested that the negative charge on the Earth is maintained by the action of thunderstorms. Thus all the stormy weather regions worldwide (on average, at any time a total of about 2,000 thunderstorms are occurring over about 10 percent of the Earth's surface) constitute the global thunderstorm generator, while the fair weather regions (about 90 percent of the globe) can be viewed as a resistive load. Lateral currents are assumed to flow freely along the highly conducting Earth's surface and in the electrosphere. The fair-weather current of the order of 1 kA must be balanced by the total generator current that is composed of currents associated with corona, precipitation, and lightning discharges. The global electric circuit concept is illustrated in Fig. 1.13. Negative charge is brought to Earth mainly by lightning discharges (most of which transport negative charge to the ground) and by corona current under thunderclouds. The net precipitation current is thought to transport positive charge to the ground. Positive

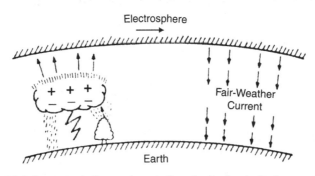

Fig. 1.13   Illustration of the global electric circuit. Shown schematically under the thundercloud are precipitation, lightning, and corona. Adapted from Pierce (1974).

charge is presumed to leak from cloud tops to the electrosphere. If we divide the potential of the electrosphere, 300 kV, by the fair-weather current, 1 kA, the effective load resistance will be 300 ohm.

Questions and problems

1.1. What is the most common type of lightning?
1.2. What is the average number of strokes in (a) negative and (b) positive flashes?
1.3. What is the typical peak current of (a) the first and (b) subsequent strokes in negative cloud-to-ground lightning?
1.4. How can lightning be triggered artificially from natural thunderclouds?
1.5. What is the difference between the upward lightning initiated from a tall tower and rocket-triggered lightning?
1.6. List transient luminous events that occur between cloud tops and the ionosphere.
1.7. What is a TGF?
1.8. Name the processes which serve to charge the capacitor formed by the electrosphere and Earth's surface? Why is this capacitor leaky?

Further reading

Appendix 1: How is lightning initiated in thunderclouds?
Appendix 4: Compact intracloud discharges (CIDs)
Cooray, V. (2015). *An Introduction to Lightning*, 386 pp., New York: Springer.
Dwyer, J. R., Smith, D. M., and Cummer, S. A. (2012). High-energy atmospheric physics: Terrestrial gamma-ray flashes and related phenomena. *Space Science Reviews* 173 (1–4), 133–196.
Rakov, V. A. and Uman, M. A. (2003). *Lightning: Physics and Effects*, 687 pp., New York: Cambridge.
Uman, M. A. (1987). *The Lightning Discharge*, 377 pp., San Diego, California: Academic Press.

2 Incidence of lightning to areas and structures

2.1 Introduction

The number of flashes per unit area per unit time is called the "flash density," and the number of ground flashes per unit area per unit time (usually per square kilometer per year) is called the "ground flash density." A global flash rate of 100 s^{-1} corresponds to a flash density over the entire Earth's surface of about $6 \text{ km}^{-2} \text{ yr}^{-1}$ and a ground (land or sea) flash density of about $1.5 \text{ km}^{-2} \text{ yr}^{-1}$ if we assume a three-to-one global ratio of cloud to cloud-to-ground flashes. Lightning activity is highly variable on both (1) a temporal scale at a given location, ranging from some tens of years to the lifetime of individual thunderstorm cells, which is of the order of one hour (Chapter 3), or shorter, and (2) a spatial scale ranging from continents and oceans to a few kilometers or less. The ground flash density N_g is the basic information needed for the estimation of the lightning strike incidence to a ground-based object or system. Other measures of the lightning incidence to an area include the annual number of thunderstorm days T_D and the annual number of thunderstorm hours T_H. The latter two measures are usually converted into N_g using empirical formulas and are generally considered to be the last resort, when no suitable measurements of N_g for the area in question are available.

2.2 Thunderstorm days

The annual number of thunderstorm days T_D, also called the "keraunic level," is the only parameter related to the lightning incidence for which worldwide data are available extending over many decades. In fact, in some specific locations such data are available for more than a century. A thunderstorm day is defined as a local calendar day during which thunder is heard at least once at a given location. The practical range of audibility of thunder is about 15 km, with the maximum range of audibility being typically about 25 km. Thunderstorm-day data are recorded at most weather stations, generally by human observers, and the spatial distribution of long-term average values of T_D is often presented in the form of isokeraunic maps. Examples of such maps are shown in Figs. 2.1 and 2.2, for the entire world and for the contiguous United States, respectively.

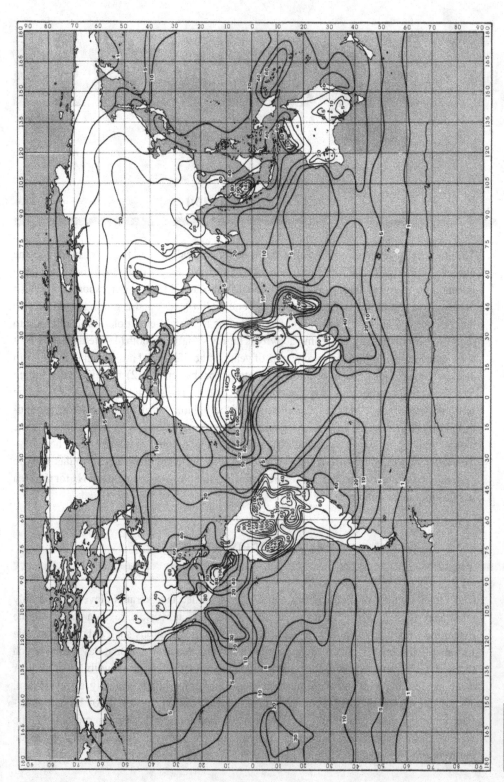

Fig. 2.1 A world map of mean annual number of thunderstorm days. Adapted from WMO Publication 21 (1956).

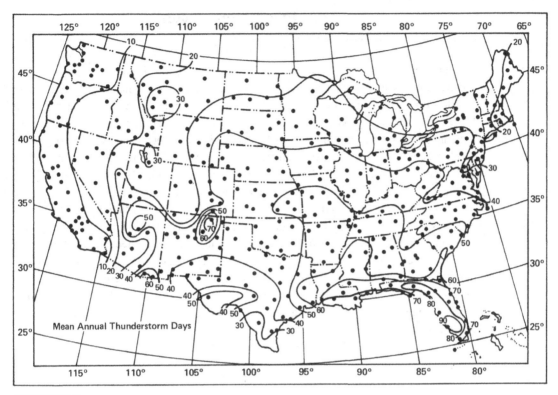

Fig. 2.2 A map of mean annual number of thunderstorm days for the contiguous United States based on data from 450 air weather stations shown as dots. Most stations had 30-year records and all had at least ten-year records. Adapted from MacGorman et al. (1984).

Considerable effort has been expended on relating the annual number of thunderstorm days T_D to the ground flash density N_g. Apparently the most reliable expression relating N_g and T_D is the following one:

$$N_g = 0.04 \, T_D^{1.25} \qquad\qquad (2.1)$$

This expression, which is found in most lightning protection standards (e.g. IEEE Std 1410–2010), is based on the regression equation relating the logarithm of the five-year-average value of N_g measured with CIGRE 10 kHz lightning flash counters at 62 locations in South Africa and the logarithm of the value of T_D as reported by the corresponding weather stations. (CIGRE is a French acronym which stands for the International Council on Large Electric Systems, and IEEE stands for the Institute of Electrical and Electronics Engineers.) The range for T_D was from 4 to 80, the range for N_g was from about 0.2 to about 13 flashes $km^{-2} \, yr^{-1}$, and the correlation coefficient between the logarithms of N_g and T_D was 0.85. For rough estimates, a simpler equation, $N_g = 0.1 \, T_D$, can be used.

2.3 Thunderstorm hours

The annual number of thunderstorm hours T_H is a parameter that is potentially more closely related to the lightning incidence than T_D. Clearly, T_D does not distinguish between a small thunderstorm producing a few lightning flashes in tens of minutes and a large storm lasting for several hours and producing hundreds of flashes, while T_H does. However, the long-term annual number of lightning-caused outages of power lines that have similar geometrical and electrical characteristics and are located in areas with different long-term values of T_D and T_H do not show a better correlation with T_H than with T_D. Similar to T_D, T_H is routinely recorded by human observers at weather stations. T_H records are perhaps more likely to suffer from human error than are T_D records. Indeed, in the case of T_D the record is just "yes" or "no" for each day, whereas in the case of T_H the observer must record the beginning and the end of storm, the end usually being defined as the time after which thunder is not heard for at least 15 minutes. If thunder is heard again after, say, 20 minutes, a new beginning is recorded. It is possible that an observer may be hesitant to regard the final audible thunder as the end of the storm when non-audible (out of thunder-hearing range) but visible lightning activity is present—a situation that is more likely to occur at night.

Perhaps the most popular relation between N_g and T_H is:

$$N_g = 0.054 \, T_H^{1.1} \tag{2.2}$$

This equation was proposed by MacGorman et al. (1984), who used ground flash data from lightning locating systems operated in Florida and Oklahoma and inferred values of T_H that a human observer would report if located at grid points with 25 km spacing. They used Eq. 2.2 to convert a map for T_H for the contiguous United States into a corresponding map for N_g. For rough estimates, a simplified version of Eq. 2.2, $N_g = 0.05 \, T_H$, can be used.

2.4 Ground flash density

The ground flash density N_g is often viewed as the primary descriptor of lightning incidence, at least in lightning protection studies. Ground flash density has been estimated from records of (1) lightning flash counters (LFCs) and (2) lightning locating systems (LLSs), and can potentially be estimated from records of satellite-based optical detectors.

Lightning Flash Counters. The lightning flash counter (LFC) is an antenna-based instrument that produces a registration if the electric (or magnetic) field generated by lightning, after being appropriately filtered (the center frequency is typically in the range from hundreds of hertz to tens of kilohertz), exceeds a fixed threshold level. The output of an LFC is the number of lightning events and/or time sequence of lightning events recorded at a given location. If the fraction of ground flashes in the total number of lightning flash counter registrations Y_g and its effective range R_g for ground flashes are known, LFCs can provide

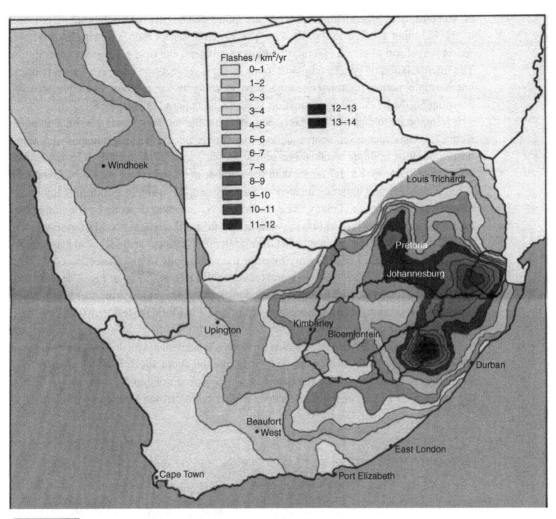

Fig. 2.3   Annual ground flash density map for South Africa based on 1975–1986 (11 years) data acquired using a network of lightning flash counters. Courtesy of R. B. Anderson.

reasonably accurate data on ground flash density. However, estimation of Y_g and R_g is not a trivial task (see Rakov and Uman, 2003, Ch. 2). As an example, a map of N_g based on LFC observations in South Africa is shown in Fig. 2.3.

Lightning Locating Systems. Locating lightning discharges with reasonable accuracy requires the use of multiple-station systems. The principles of operation of multiple-station lightning locating systems (LLSs) are described in Chapter 8. Such systems are presently used in many countries to acquire lightning data that can be used for mapping N_g. Any LLS fails to detect relatively small cloud-to-ground flashes (particularly near and outside the perimeter of the network) and fails to discriminate against some cloud flashes that are unwanted in determining N_g. The corresponding system characteristics, the detection efficiency and the classification accuracy, are influenced by network configuration, position of the lightning relative to the network, the system's sensor gain and trigger threshold, sensor waveform selection criteria, lightning parameters, and field propagation conditions. The interpretation of system output in terms of N_g is subject to a number of uncertainties, but multiple-station lightning locating networks are by far the best available tool for mapping N_g. More detailed information about LLSs is found in Chapter 8.

It is important to note that LLSs record strokes, not flashes, and therefore estimation of N_g from LLS data depends on the method to group strokes into flashes. Further, many lightning flashes produce multiple terminations on the ground, so that the number of ground strike points is a factor of 1.5–1.7 larger than the number of flashes. This should be taken into account in estimating lightning incidence to areas when, for example, performing lightning-related risk calculations. Finally, the accuracy of N_g mapping depends on the number of events per grid cell, which in turn depends on the grid cell size and period of observations. IEEE Std 1410–2010 recommends that the number of events per grid cell be at least 400.

An example ground flash density map for the contiguous United States, based on data from the National Lightning Detection Network (NLDN), is shown in Fig. 2.4. It is clear from this map that variation in ground flash density from one region to another is more than two orders of magnitude (from more than $12 \text{ km}^{-2} \text{ yr}^{-1}$ in Florida to less than $0.25 \text{ km}^{-2} \text{ yr}^{-1}$ along the Pacific coast).

Satellite Detectors. Optical satellite detectors cannot distinguish between cloud and ground discharges and, hence, provide information about the total flash density N_{tot} (including both cloud and ground discharges). In order to obtain N_g from a map of N_{tot} (an example is shown in Fig. 2.5), a spatial distribution of the fraction of discharges to

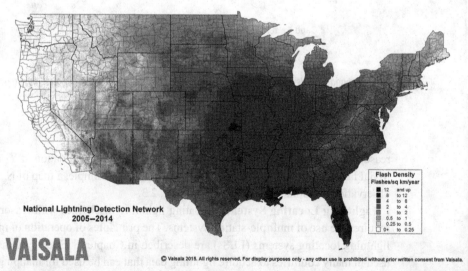

National Lightning Detection Network
2005–2014

Flash Density
Flashes/sq km/year
■ 12   and up
■ 8   to 12
■ 4   to 8
■ 2   to 4
■ 1   to 2
■ 0.5 to 1
□ 0.25 to 0.5
□ 0+   to 0.25

VAISALA

© Vaisala 2015. All rights reserved. For display purposes only - any other use is prohibited without prior written consent from Vaisala.

Fig. 2.4 Annual ground flash density map (2 km^2 grid) for the contiguous United States based on 2005–2014 (10 years) NLDN data. Courtesy of Vaisala, Inc. Color version of this map is found at http://www.vaisala.com/VaisalaImages/ avg_fd_2005–2014_CONUS_2km_grid.png

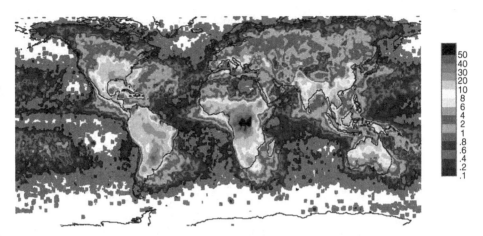

Fig. 2.5 A global map of total lightning flash density (per square kilometer per year) based on data from two satellite detectors, Optical Transient Detector (OTD, five years) and Lightning Imaging Sensor (LIS, three years). Grey areas: 0.01–0.1 $km^{-2}yr^{-1}$; white areas: <0.01 $km^{-2}yr^{-1}$. Courtesy of H. J. Christian, NASA/Marshall Space Flight Center.

ground relative to the total number of lightning discharges is needed. The global average of this fraction is 0.25. IEEE Std 1410–2010 recommends, in the absence of ground-based measurements of N_g, using the total flash density based on satellite observations and assuming the fraction of ground flashes to be one-third.

2.5 Lightning incidence to various objects

2.5.1 General information

We first briefly describe how cloud-to-ground lightning "decides" on its ground termination point. More details are given in Chapters 4 and 9. Ground flashes are normally initiated by stepped leaders that originate in the thundercloud. As the downward-extending leader channel, usually negatively charged, approaches the ground, the enhanced electric field intensity at irregularities of the Earth's surface or at protruding grounded objects increases and eventually exceeds the breakdown value of air. As a result, one or more upward-moving leaders are initiated from those points. When one of the upward-moving leaders from the ground contacts a branch of the downward-moving stepped leader, the return stroke is initiated. Grounded vertical objects produce relatively large electric field enhancement near their upper extremities so that upward-moving connecting leaders from these objects start earlier than from the surrounding ground and, therefore, serve to make the object a preferential lightning termination point. In general, the higher the object, the greater the field enhancement and hence the higher the probability that a stepped leader will terminate on the object. At the limit, when the height (field enhancement capability, to be more exact) of the object becomes so large that the upward-moving leader from the object tip can be initiated by in-cloud charges, in-cloud discharge processes, or relatively distant CG or IC flashes – as opposed to being initiated by the charge on the approaching stepped leader – the object becomes capable of initiating upward lightning, as discussed in Chapter 1. The latter, as opposed to "normal," downward lightning, would not occur if the object were not there. Ground-based objects with heights ranging from about 100 to 500 m experience both downward and upward flashes, with the proportion being a function of object height.

Eriksson (1987) derived the following equation for the annual lightning incidence N (in yr^{-1}) to ground-based objects, including both downward and upward (if any) flashes:

$$N = 24 \times 10^{-6}\, H_s^{2.05}\, N_g \qquad\qquad (2.3)$$

where H_s is the object height in meters and N_g is in $\text{km}^{-2}\,\text{yr}^{-1}$. To do so, he employed (1) the observations of lightning incidence to structures of heights ranging from 20 to 540 m in different countries, (2) the corresponding local values of the annual number of thunderstorm days T_D, and (3) Eq. 2.1. It is worth noting that the majority of observed lightning strikes involved in the derivation of Eq. 2.3 correspond to taller structures, and that the observed lightning incidence to smaller-height structures might have been affected by the presence of surrounding objects such as buildings and trees. As a result, Eq. 2.3 is probably less reliable for H_s less than 60 m or so than for greater heights. Eriksson (1978) tabulated the observed percentage of upward flashes as a function of a free-standing structure's height, as reproduced in Table 2.1. Eriksson and Meal (1984) fitted the data in Table 2.1 with the following expression:

$$P_u = 52.8 \ln H_s - 230 \qquad\qquad (2.4)$$

where P_u is the percentage of upward flashes, H_s is the structure height in meters, and ln is the natural (base-e) logarithm. This equation is valid only for structure heights ranging from 78 to 518 m, since for $H_s = 78$, m $P_u = 0$ and for $H_s = 518$ m, $P_u = 100\%$. Structures with heights less than 78 m are not covered by Eq. 2.4 because they are expected to be struck by downward flashes only, and structures with heights greater than 518 m are not covered because they are generally expected to experience upward flashes only. In practice, as stated above, structures having heights less than 100 m or so are often assumed to be struck by downward lightning only, and the upper height limit can be simply taken as 500 m. Accordingly, the total lightning incidence N to a structure is the sum of the downward-flash incidence N_d and upward-flash incidence N_u if the structure height is in the range from about 100 to 500 m, $N = N_d$ for structures shorter than 100 m, and $N = N_u$ for structures taller than 500 m. There are, however, some recently reported exceptions to this simple rule. For example, the 634 m-high Tokyo Sky Tree in Japan experiences a significant number of unexpected downward flashes and, as of this writing, only downward flashes were observed striking high-rise (>300 m) buildings in Guangzhou, China (e.g., Lu et al., 2013).

If both downward and upward flashes are expected, they are often treated separately in estimating the lightning incidence to an object, as described below.

Table 2.1 The percentage of upward flashes initiated from tall structures. Adapted from Eriksson (1978)

Reference	Structure height, m	Percentage of upward flashes
Pierce (1972)	150	23
	200	50
	300	80
	400	91
McCann (1944)	110	8
	180	24
	400	96
Berger (1972)	350[a]	84
Gorin (1972); Gorin et al. (1976)	540	92[b]
Garbagnati et al. (1974)	500[c]	98

[a] An effective height of 350 m has been assigned by Eriksson to Berger's 70 m high mountaintop towers to account for the enhancement of the electric field by the mountain whose top is 640 m above Lake Lugano (914 m above sea level).
[b] 50 percent of the flashes recorded in this study were classified as "unidentified." The relative incidence of upward flashes is based upon analysis of only the identified data.
[c] An effective height of 500 m has been assigned by Eriksson to Garbagnati et al.'s towers that were 40 m high and located on mountaintops, 980 and 993 m above sea level.

2.5.2 Downward flashes

When the incidence of downward lightning is estimated, it is common to ascribe a so-called equivalent attractive (or exposure) area to the grounded object. The attractive area can be viewed as an area on the ground surface that would receive the same number of lightning strikes in the absence of the object as does the object placed in the center of that area. In other words, in computing lightning incidence to a structure, the structure is replaced by an equivalent area on the ground. For a free-standing structure whose plan-view dimensions are much smaller than its height (such as a mast, tower, or chimney), this area, A, is circular and is generally given by $A = \pi R_a^2$, where R_a is the equivalent attractive radius, discussed later. For straight, horizontally extended structures (such as power lines or their sections; see Fig. 2.6), the equivalent attractive area is rectangular and is sometimes termed the "shadow zone" or "attractive swath." For example, if a structure has a length l and an effective width b, its equivalent attractive area is generally estimated as $A = 1 \times W = 1(b + 2R_a)$, where R_a is the equivalent attractive distance generally thought to be approximately equal to the equivalent attractive radius for a free-standing structure of the same height (e.g. Eriksson,

Fig. 2.6   Power transmission lines, each having three phase conductors (labeled PC), suspended on metallic towers using insulators, and two overhead ground wires (labeled OGW), installed above phase conductors for lightning protection. Courtesy of Mannvit Engineering (http://cavanbi1971.pixnet.net/blog/post/87292142-download-trans mission-towers)

1987). For power transmission lines, the effective width b (see Fig. 2.7a) is usually taken as the horizontal distance between overhead ground wires (b = 0 for the case of a single ground wire). Further, the local ground flash density N_g is assumed to be spatially uniform in the absence of the structure, so that the downward lightning incidence to the structure is found as

$$N_d = AN_g \qquad (2.5)$$

Usually N_g is in $km^{-2} yr^{-1}$ so that A should be expressed in km^2 to obtain N_d in yr^{-1} (strikes per year).

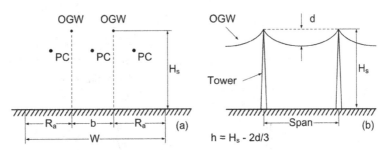

Fig. 2.7 (a) Schematic representation of power transmission line having three phase conductors (labeled PC) and two overhead ground wires (labeled OGW), the latter being separated by distance b. Also shown is the attractive distance R_a and the width W of equivalent attractive area A of the line. (b) Illustration of the OGW mid-span sag d, which is used for estimating the effective height h of the line. Artwork by Potao Sun.

The equivalent attractive radius (or distance) R_a is usually assumed to be a function of structure height H_s and is generally expressed as

$$R_a = \alpha \, H_s^\beta \qquad (2.6)$$

where α and β are empirical constants found using observed lightning incidence (annual number of lightning strikes) to structures of different height and estimated values of N_g. In Eq. 2.6, both H_s and R_a are in meters, and different values of α and β have been proposed. According to the IEEE Std 1243–1997, $\alpha = 14$ and $\beta = 0.6$ and the corresponding power-line flash collection rate, N_s (in strikes per 100 km per year), is

$$N_s = N_g(28 \, h^{0.6} + b)/10. \qquad (2.7)$$

In Eq. 2.7, h is the effective height of the line taking into account conductor (usually overhead ground wire) sags. It is found as $h = H_s - 2d/3$, where H_s is the height of power-line towers and d is the mid-span sag (see Fig. 2.7b). Both h and b are in meters, and N_g is in $km^{-2} \, yr^{-1}$. For rough estimates the attractive radius (distance) can be expressed as $R_a = 2H_s$.

The attractive radius for individual strikes should depend on the charge carried by the descending leader, this charge being correlated with the associated return-stroke peak current (Chapter 9). In this regard, Eq. 2.6 should be understood as representing the entire distribution of peak currents. In the so-called electrogeometrical approach (Chapter 9), which is widely used for the estimation of lightning incidence in lightning protection studies, the equivalent attractive radius explicitly depends on the statistical distribution of lightning peak currents (e.g. Eriksson, 1987).

Estimation of N_d from Eq. 2.5 implies a reasonably long-term value of ground flash density and yields a long-term average value of lightning incidence. For example, if a 60 m tower is located in a part of Florida where $N_g = 10 \ km^{-2} \ yr^{-1}$, the long-term average downward lightning incidence will be about $0.5 \ yr^{-1}$ (assuming $\alpha = 2$ and $\beta = 1$); that is, the tower will be struck on average every other year. The use of Eq. 2.3 would result in a lightning incidence value of about $1 \ yr^{-1}$. For a house (or other similar structure) located in

Table 2.2 Number, $N = 200 \, p(n)$, of houses out of a total of 200 (percentage if divided by 2) expected to be struck by lightning n times over T years ($N_g = 4 \text{ km}^{-2} \text{ yr}^{-1}$; $A = 1200 \text{ m}^2$)

Number of years of observation, T	Number of times struck by lightning, n				
	0	1	2	3	>3
10	191	9	0	0	0
20	182	17	1	0	0
30	173	25	2	0	0
40	165	32	3	0	0
50	157	38	5	0	0
60	150	43	6	1	0
70	143	48	8	1	0
80	136	53	10	1	0
90	130	56	12	2	0
100	124	60	14	2	0

a region characterized by a moderate ground flash density of $4 \text{ km}^{-2} \text{ yr}^{-1}$ and having an area of $10 \times 20 \text{ m}^2$ and a height of 5 m so that the equivalent attractive distance is about 10 m (extrapolating Eq. 2.6 with $\alpha = 2$ and $\beta = 1$ to a structure height of 5 m), the approximate equivalent attractive area is $30 \times 40 = 1200 \text{ m}^2$. Such a house is expected to be struck by lightning $(1200 \text{ m}^2) \, (10^{-6} \text{ km}^2 \text{ m}^{-2}) \, (4 \text{ km}^{-2} \text{ yr}^{-1}) = 4.8 \times 10^{-3}$ times a year, or about once every 200 years on average. Another way to think of this lightning incidence is that, in this region, one of every 200 houses will be struck each year, on average.

The probability of a structure, represented by its equivalent attractive area A, to be struck exactly 0, 1, 2, 3, ... n times in T years can be estimated using the Poisson probability distribution,

$$p(n) = (Z^n/n!)/\exp(-Z) \tag{2.8}$$

where $Z = AN_gT = N_d \, T$, the average number of strikes expected over T years, provided that N_g remains constant. Continuing the previous example, Table 2.2 gives the number of houses, out of a total of 200, expected to be struck by lightning n times over T years, as predicted by Eq. 2.8. Each number is found as the product of the total number of houses, 200, and $p(n)$ from Eq. 2.8 with subsequent rounding off to the nearest integer. In two cases ($n = 1$; $T = 80 \text{ yr}$ and $T = 100 \text{ yr}$) the rounded-off number was increased by 1 in order to assure that the sum of numbers in each row is 200, the total number of houses considered. It follows from Table 2.2 that, for example, over a 60-year period 150 houses will not be struck at all, 50 $(43 + 6 + 1)$ will receive at least one strike, six houses will be struck twice, and one house will be struck three times. If $N_g = 12 \text{ km}^{-2} \text{ yr}^{-1}$, characteristic of some areas in Florida, then over a period of 60 years (perhaps the lifetime of a house) only 84 houses out of the 200 will receive no lightning strikes, and 11 houses will be struck three times or more.

2.5.3 Upward flashes

Once the incidence of downward lightning N_d is found from Eq. 2.5 using the concept of an equivalent attractive area, the incidence of upward flashes N_u can be determined by subtracting N_d from N given by Eq. 2.3. Recall that if the structure height is less than 100 m or so, it is usually assumed that $N_u = 0$. If only the percentage of upward flashes is sought, Eq. 2.4 can be used.

2.6 Summary

The global lightning flash rate is some tens to 100 per second or so. The most common measures of lightning incidence to an area are the annual number of thunderstorm days, the annual number of thunderstorm hours, and the annual ground flash density. Annual ground flash density is the primary descriptor. It has been estimated from records of lightning flash counters and lightning locating systems. The observed variation in ground flash density from one region to another in the United States is one to two orders of magnitude. Many flashes strike the ground at more than one point. When only one location per flash is recorded, the correction factor for measured values of ground flash density, in order to take into account multiple channel terminations on the ground, is 1.5 to 1.7. Ground-based objects with heights ranging from about 100 to 500 m experience both downward and upward flashes, with the proportion being a function of the height of the object. Structures having heights less than 100 m or so are often assumed to be struck only by downward lightning, while those with heights greater than 500 m or so are assumed to experience only upward flashes. A house located in a region characterized by a moderate ground flash density of $4 \, \text{km}^{-2} \, \text{yr}^{-1}$ and having an area of $10 \times 20 \, \text{m}^2$ and a height of 5 m is expected to be struck by lightning roughly once every 200 years.

Questions and problems

2.1. Name and compare three descriptors of lightning incidence to an area.
2.2. What is the lightning flash counter? How can it be used for estimating the ground flash density?
2.3. Can the ground flash density be estimated using data from optical satellite detectors?
2.4. Which part of (a) the world and (b) the United States experiences the highest level of lightning activity?
2.5. A tower 175 m high is located near Tampa, Florida. Estimate the annual lightning incidence to the tower and the percentage of upward flashes initiated from the tower.
2.6. Find the annual lightning incidence to a single-circuit power transmission line with two overhead ground wires (see Figs. 2.6 and 2.7), whose length is 200 km. Assume that $H_s = 39$ m, b = 16 m, d = 12 m, and that the annual number of thunderstorm days is 40.

2.7. A 60 m tower is struck by lightning on average once every four years. Estimate the average ground flash density in the area where the tower is located.

2.8. For the tower in Problem 7, find the probability of being struck by lightning exactly 0, 1, and 2 times during (a) 2 years and (b) 10 years.

Further reading

Appendix 5: Is it true that lightning never strikes the same place twice?

Rakov, V. A. and Uman, M. A. (2003). *Lightning: Physics and Effects*, 687 pp., New York: Cambridge University Press.

Uman, M. A. (1987). *The Lightning Discharge*, 377 pp., San Diego, California: Academic Press.

3 Electrical structure of thunderclouds

3.1 General information about thunderstorms

The primary source of lightning is the cloud type termed "cumulonimbus," commonly referred to as the "thundercloud." Lightning-like electrical discharges can also be generated in the ejected material above volcanoes, in sandstorms, and in nuclear explosions.

Before reviewing the electrical structure of thunderclouds it is worth outlining their meteorological characteristics. In effect, thunderclouds are large atmospheric heat engines with the input energy coming from the Sun and with water vapor as the primary heat-transfer agent (Moore and Vonnegut, 1977). The principal outputs of such an engine include (but are not limited to) (1) the mechanical work of the vertical and horizontal winds produced by the storm, (2) an outflow of condensate in the form of rain and hail from the bottom of the cloud and of small ice crystals from the top of the cloud, and (3) electrical discharges inside, below, and above the cloud, including corona, lightning, sprites, halos, elves, blue starters, blue jets, and gigantic jets. The processes that operate in a thundercloud to produce these actions are many and complex, most of them being poorly understood. A thundercloud develops from a small, fair-weather cloud called a "cumulus," which is formed when parcels of warm, moist air rise and cool by adiabatic expansion; that is, without the transfer of heat or mass across the boundaries of the air parcels. When the relative humidity in the rising and cooling parcel exceeds saturation, moisture condenses on airborne particulate matter to form the many small water particles that constitute the visible cloud. The height of the condensation level, which determines the height of the visible cloud base, increases with decreasing relative humidity at ground level. This is why cloud bases in Florida are generally lower than in arid locations such as New Mexico or Arizona. Parcels of warm, moist air can only continue to rise to form a cumulus and eventually a cumulonimbus if the atmospheric temperature lapse rate – the decrease in the temperature with increasing height – is larger than the moist-adiabatic lapse rate of about 0.6°C per 100 m. The atmosphere is then referred to as unstable since rising moist parcels remain warmer than the air around them and thus remain buoyant. When a parcel rises above the 0°C isotherm, some of the water particles begin to freeze, but others (typically smaller particles) remain liquid at temperatures colder than 0°C. These are called supercooled water particles. At temperatures colder than about −40°C all water particles will be frozen. In the temperature range from 0°C to −40°C liquid water and ice particles coexist, forming a mixed-phase region where most electrification is thought to occur (Section 3.4).

Convection of buoyant moist air is usually confined to the troposphere, the layer of the atmosphere that extends from the Earth's surface to the tropopause. The latter is the boundary between the troposphere and the stratosphere, the layer which extends from the tropopause to a height of approximately 50 km. In the troposphere the temperature decreases with increasing altitude, while in the stratosphere the temperature at first becomes roughly independent of altitude and then increases with altitude. A zero or positive temperature gradient in the stratosphere serves to suppress convection and, therefore, hampers the penetration of cloud tops into the stratosphere. The height of the tropopause varies from approximately 18 km in the tropics in the summer to 8 km or so in high latitudes in the winter. In the case of vigorous updrafts, cloud vertical growth continues into the lower portion of the stratosphere. Convective surges can overshoot the tropopause by up to 5 km in severe storms.

Although the primary thunderstorm activity occurs in the lower latitudes, thunderclouds are occasionally observed in the polar regions. Thunderstorms commonly occur over warm coastal regions when breezes from the water are induced to flow inland after sunrise when the land surface is warmed by solar radiation to a temperature higher than that of the water. Similarly, because mountains are heated before valleys, they often aid the onset of convection in unstable air. Further, horizontal wind blowing against a mountain will be directed upward and can aid in the vertical convection of air parcels, a process which is referred to as the "orographic effect." While relatively small-scale convective thunderstorms (also called air-mass thunderstorms) develop in the spring and summer months, when the potential for convection is usually the greatest and adequate water vapor is available, larger-scale storms associated with frontal activity are observed in temperate latitudes at all times throughout the year. Frontal thunderstorms are formed when, for example, a relatively large mass of cold air moves southward over the United States from the high latitudes and slides under warmer moister air.

The horizontal dimensions of active air-mass thunderstorms range from about 3 km to greater than 50 km. Seemingly merged thunderclouds may occur in lines along cold fronts extending for hundreds of kilometers. Ordinary thunderstorms can be viewed as being composed of units of convection, typically some kilometers in diameter, characterized by relatively strong updrafts (≥ 10 m s^{-1}). These units of convection are referred to as "cells." The lifetime of an individual cell is of the order of one hour. Thunderstorms can include a single isolated cell, several cells, or a long-lived cell with a rotating updraft, called a "supercell." At any given time a typical multicell storm consists of a succession of cells at different stages of evolution. Large frontal systems have been observed to persist for more than 48 hours and to move over more than 2,000 km. Thunderstorms over flat terrain tend to move at an average speed of 20 to 30 km hr^{-1}. Further information on thunderstorm morphology and evolution can be found in the book by MacGorman and Rust (1998) and in the book chapter by Williams (1995).

3.2 Idealized gross charge distribution

3.2.1 General information

The distribution and motion of thunderstorm electric charges, most residing on hydrometeors (various liquid or frozen water particles in the atmosphere) but with some free ions, is complex and changes continuously as the cloud evolves. Hydrometeors whose motion is predominantly influenced by gravity (fall speed ≥ 0.3 m s^{-1}) are called precipitation. All other hydrometeors are called cloud particles. The basic features of the cloud charge structure include a net positive charge near the top, a net negative charge below it, and an additional positive charge at the bottom of the cloud. These features appear to be generally accepted and are illustrated (along with the negative screening layer charge at the cloud top and positive corona space charge at ground) in Fig. 3.1. Note that the lower positive charge is depicted in Fig. 3.1 as carried by precipitation. The lower positive charge region is further discussed in Section 3.5.

3.2.2 Simple model

In computing cloud electric fields or lightning-caused field changes, the charge structure shown in Fig. 3.1 is often approximated by three vertically stacked point charges (or

Fig. 3.1 An isolated thundercloud in central New Mexico and a rudimentary picture of how electric charge is thought to be distributed inside and around the thundercloud, as inferred from the remote and in-situ observations. Adapted from Krehbiel (1986).

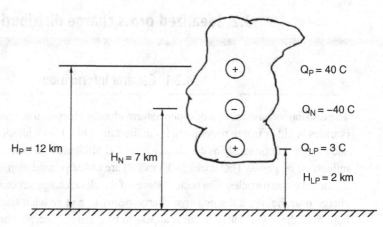

Fig. 3.2a A vertical tripole representing the idealized gross charge structure of a thundercloud such as that shown in Fig. 3.1 with the negative screening layer charges at the cloud top and the positive corona space charge produced at ground being ignored here. © Vladimir A. Rakov and Martin A. Uman 2003, published by Cambridge University Press, reprinted with permission.

spherically symmetrical charged volumes). This charge configuration includes a positive charge at the top, negative in the middle, and an additional, smaller positive at the bottom, all located in a non-conducting atmosphere above a perfectly conducting ground, as illustrated in Fig. 3.2a. The top two charges are usually called the "main" charges and are often specified to be equal in magnitude. It is thought that the lower positive charge may not always be present. The two main charges form a dipole, said to be positive because the positive charge is above the negative (giving an upward-directed dipole moment).

3.2.3 Electric field due to cloud charges

The electric field intensity $\overline{E}$ due to the system of three charges shown in Fig. 3.2a is found by replacing the perfectly conducting ground with three image charges and using the principle of superposition, with the total electric field being the vectorial sum of six contributions (three from the actual charges and three from their images). The computation of the electric field on the ground surface due to the main negative charge and its image is illustrated, with reference to Fig. 3.2b, below. Note that in this illustration, including Fig. 3.2b, the subscript "N" is dropped for simplicity and the charge signs are shown explicitly. Due to the symmetry of the problem with respect to the field point P, the magnitudes of the contributions from the actual charge and its image are equal, and each contribution is found as

$$|\overline{E}^{(-)}| = |\overline{E}^{(+)}| = \frac{|Q|}{4\pi\varepsilon_0(H^2 + r^2)} \tag{3.1}$$

It is clear from Fig. 3.2b that the components of the electric field tangential to the ground plane due to the actual charge and its image cancel each other, as expected from the

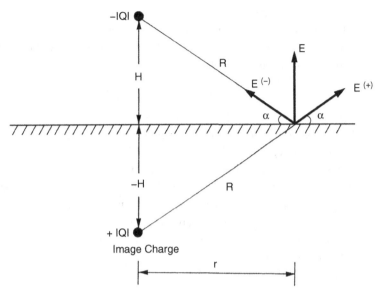

Fig. 3.2b   Illustration of the calculation of the electric field due to a negative point charge above a perfectly conducting ground for a field point located at the ground surface. © Vladimir A. Rakov and Martin A. Uman 2003, published by Cambridge University Press, reprinted with permission.

boundary condition on the surface of a perfect conductor. The electric field components normal to the ground plane add, with the total normal field magnitude being twice the contribution from either the actual charge or its image:

$$
\begin{aligned}
\left|\overline{E}\right| &= 2\left|\overline{E}^{(-)}\right|\cos(90° - \alpha) = 2\left|\overline{E}^{(+)}\right|\cos(90° - \alpha) \\
&= 2\left|\overline{E}^{(-)}\right|\sin\alpha = \frac{|Q|H}{2\pi\varepsilon_0(H^2 + r^2)^{3/2}}
\end{aligned}
\tag{3.2}
$$

To facilitate further discussion of the variation of $|\overline{E}|$ as a function of H and r ($|Q|$ = const), we re-write Eq. 3.2 in the following form:

$$
\left|\overline{E}\right| = k\frac{\sin\alpha}{R^2}
\tag{3.3}
$$

where $k = |Q|/(2\pi\varepsilon_0)$ and $R^2 = (H^2 + r^2)$. For a fixed r, $|\overline{E}|$ increases as H increases from zero to $H = r/\sqrt{2}$ (because $\sin\alpha$ increases faster than R^2 does) and then decreases as H increases further (because R^2 increases faster than $\sin\alpha$ does). The magnitude of electric field is zero when $\alpha = 0$ and maximum when $\alpha = 35.3°$. If we now fix H and vary r, $|\overline{E}|$ will decrease monotonically with increasing r. The rate of decrease depends on H, being slower for larger H. As a result, for two vertically stacked charges of equal magnitude (say, the main positive and main negative charges in Fig. 3.2a), the relative contribution to $\overline{E}$ from each of these two charges will depend on r. The electric field at r = 0 is dominated by the lower charge (since it is closer and $\sin\alpha$ is the same for both charges), but as r increases, so does the relative contribution from the upper charge. At a certain distance, the contribution from the upper charge becomes dominant, and the total electric field (the sum of the contributions from the two charges) changes its polarity. The distance at which the two contributions are equal to each other (add to zero) is called the "reversal distance." For the case of two vertically-stacked charges of equal magnitude but opposite polarity, the reversal distance, D_0, is given by

$$D_0 = [(H_P H_N)^{2/3}(H_P^{2/3} + H_N^{2/3})]^{1/2} \tag{3.4}$$

where H_P and H_N are the heights of the positive and negative charges, respectively. Thus, for a positive dipole, one might expect that at close ranges the total electric field is negative (the closer negative charge is more "visible"), while at far ranges it is positive (the larger-elevation-angle positive charge is more "visible").

The electric field at ground level due to the system of three charges (a vertical tripole) shown in Fig. 3.2a, computed assuming that the middle negative and top positive charges are 7 and 12 km above ground, respectively, each having a magnitude of 40 C, and that the bottom positive charge is at 2 km and has a magnitude of 3 C, is shown in Fig. 3.2c. An upward-directed electric field in this chapter is defined as positive (the so-called physics sign convention, which corresponds to a coordinate system with the vertical axis directed

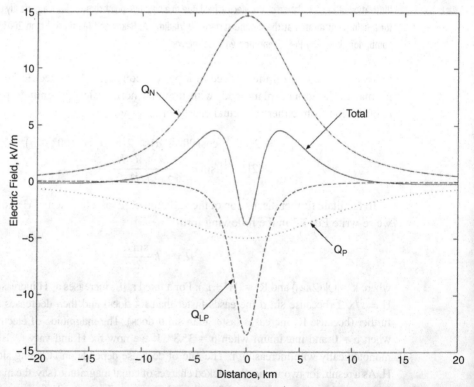

Fig. 3.2c Electric field at ground due to the vertical tripole shown in Fig. 3.2a, labeled Total, as a function of distance r from the axis (at r = 0) of the tripole. Also shown are the contributions to the total electric field from the three individual charges of the tripole. An upward directed electric field is defined as positive (physics sign convention). © Vladimir A. Rakov and Martin A. Uman 2003, published by Cambridge University Press, reprinted with permission.

upward). As seen in Fig. 3.2c, the total electric field of three vertically stacked charges exhibits two polarity reversals at the ground (neither of these reversal distances is described by Eq. 3.4 which is derived just for two vertically stacked charges). Such an electric field versus distance variation, although model-dependent, is qualitatively consistent with the available experimental data. Also shown in Fig. 3.2c are the contributions to the total electric field from each of the three charges.

3.2.4 Electric field changes due to lightning

In general, an electric field change is the difference between the final electric field value (after the charge removal due to lightning) and the initial electric field value due to the original cloud charge distribution. For any charge removed from the cloud, the corresponding electrostatic field change is the negative of the contribution of that charge to the initial electric field. If we assume that the negative cloud charge is completely neutralized as a result of a cloud-to-ground discharge, the resultant net

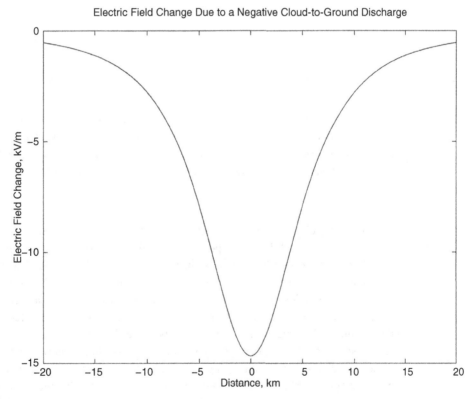

Fig. 3.2d Electric field change ΔE at the ground due to the total removal of the negative charge (Q_N) of the vertical tripole shown in Fig. 3.2a via a cloud-to-ground discharge as a function of distance from the axis of the tripole. Note that the electric field change at all distances is negative (it is the negative of the curve labeled "Q_N" in Fig. 3.2c).
© Vladimir A. Rakov and Martin A. Uman 2003, published by Cambridge University Press, reprinted with permission.

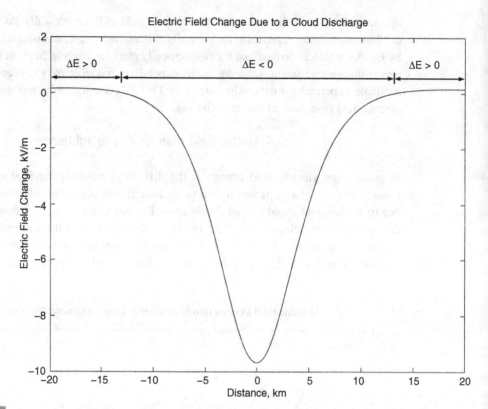

Electric Field Change Due to a Cloud Discharge

electric field change will be negative at any distance, as shown in Fig. 3.2d, because the upward-directed (positive) electric field due to the negative charge (see Fig. 3.2b) disappears; that is, becomes zero. If both main positive and main negative charges are neutralized via an intracloud discharge, the resultant net field change as a function of distance will exhibit a polarity reversal, as seen in Fig. 3.2e. Note that positive field values in Fig. 3.2e are considerably smaller than negative field values. The polarity reversal occurs because the net field change is the negative of the sum of the contributions to the total electric field, shown in Fig. 3.2c, from these two charges, this sum being positive at close ranges (dominated by the lower negative charge) and negative at far ranges (dominated by the higher positive charge). In other words, for such an intracloud discharge, the electric field change at close ranges is dominated by the reduction of the positive (upward-directed) electric field and at far ranges by the reduction of the negative (downward-directed) electric field.

3.3 Observations

The gross charge structure shown in Fig. 3.2a is based on remote (outside the cloud) and in situ (inside the cloud) measurements of electric fields. Remote measurements have been made by different researchers both at ground level and over the cloud tops using aircraft or balloons.

3.3.1 Remote measurements

Remote electric field measurements fall in two categories: (a) measurements of the slowly varying fields associated with the cloud charges or (b) measurements of the more rapid field changes associated with the neutralization of a portion of those cloud charges by lightning. Figure 3.3 shows an electric field record that illustrates both the slowly varying cloud charge electric field and the transient lightning electric field changes for a thunderstorm that grows and dies at a more or less fixed distance of about 5 km from the observation point at ground.

In clear-sky (fair-weather) conditions, the electric field vector points downward and is defined in this chapter as negative (physics sign convention). The sources of the fair-weather electric field are positive space charge in the atmosphere and negative charge on the Earth's surface. The fair-weather field has a magnitude of about 100 V m^{-1}. Beneath an active thundercloud the electric field at the ground is usually reversed in sign (directed upward) with respect to the fair-weather field and is considerably larger: 1 to 10 kV m^{-1} on relatively flat terrain. In Fig. 3.3, the fair-weather field is measured from 12:05 to about 12:30 and again after about 13:28. A large, predominantly upward-directed electric field that is indicative of a dominant negative charge in the cloud overhead is seen from roughly 12:43 to 13:08.

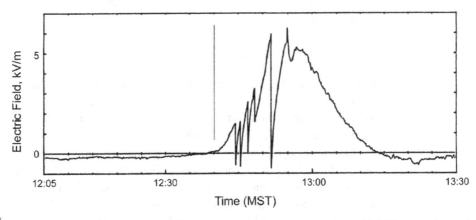

Fig. 3.3 Electric field at the ground about 5 km from a small storm near Langmuir Laboratory, New Mexico, on August 3, 1984. An upward-directed electric field is defined as positive (physics sign convention). Large pulses superimposed on the rising portion of the overall electric field waveform are due to lightning. Adapted from Krehbiel (1986).

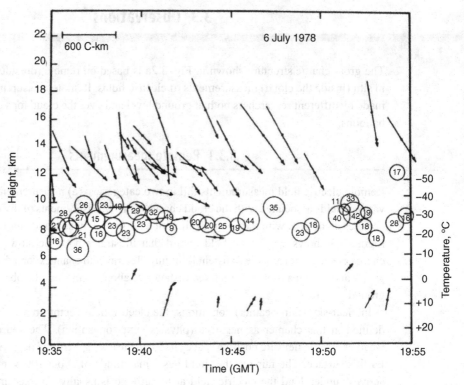

Fig. 3.4 The locations of negative charges (circles) neutralized by ground flashes and point-dipole charge moments (arrows) describing the effective positive charge transfer by cloud flashes as a function of time for a portion of an active Florida storm. Numbers in the circles give the magnitudes of neutralized charges in coulombs. In cloud flashes, negative charge was effectively transported in the direction opposite to that of the arrow to neutralize the positive charge of equal magnitude. The dot in the middle of each arrow represents the apparent single location of these two charges, the actual locations being indeterminate in the point-dipole solution (see Appendix 2). Adapted from Koshak and Krider (1989).

Koshak and Krider (1989) inferred cloud charges neutralized by lightning from lightning electric field changes recorded by the Kennedy Space Center (KSC) electric field mill network. Their results, for a portion of an active storm, are shown in Fig. 3.4. The circles represent the locations of point charge solutions that indicate where cloud-to-ground flashes remove negative charge from – or, equivalently, deposit positive charge to – the cloud. The numbers in the circles give the magnitudes of those charges in coulombs. The arrows show the locations (dot in the middle of the arrow), directions, and magnitudes of point-dipole charge moments for cloud discharges, with the directions indicating the positive charge transport. The relative positions of the circles and the majority of the arrows (above the circles) suggest that a negative charge region is located below a positive charge region. Further, the predominantly upward directed arrows below the circles are indicative of a small pocket of positive charge below the negative charge region.

3.3.2 In-situ measurements

In-situ measurements of electric field inside the cloud have been made using free balloons, aircraft, rockets, and parachuted electric field mills. Aircraft measurements are usually made at only a few selected altitudes, while balloons provide a vertical profile without systematic measurements in the horizontal direction. Measurements made with vertically ascending balloons provided the most revealing data on the electrical properties of thunderclouds.

In-situ measurements are superior to remote measurements (Section 3.3.1) in that a relatively accurate charge height can be determined. However, since the balloon can sense the field only along a more or less straight vertical path, and it samples different portions of that path at different times, the charge magnitude can be estimated only if assumptions regarding the size and shape of individual charge regions and the charge variation with time are made. The average volume charge density, ρ_V, in the cloud is generally found by assuming that the charge (1) is horizontally uniform (i.e. there is a negligible variation with x or y compared to the variation with z), which can be viewed as an assumption of charge layers of infinitely large horizontal extent, and (2) does not vary in time. Then, using Gauss' law in point form, $\nabla \cdot \overline{E} = \partial E_x/\partial x + \partial E_y/\partial y + \partial E_z/\partial z = \rho_V/\varepsilon_0$ (e.g. Sadiku, 1994), and keeping only the last term, we get $\rho_V = \varepsilon_0 \, (dE_z/dz) \approx \varepsilon_0 \, (\Delta E_z/\Delta z)$. Thus, ρ_V is proportional to the rate at which the vertical electric field E_z increases (positive charge density) or decreases (negative charge density) with increasing altitude z as the balloon ascends.

An example of measured electric field profile for a small Alabama thunderstorm, from which four charge layers were inferred using the one-dimensional approximation to Gauss' law, is presented in Fig. 3.5.

3.3.3 Main results

It has been inferred from a combination of remote and in-situ measurements that in very different environments negative charge is typically found in the same relatively narrow temperature range, roughly -10 to $-25°C$, where the clouds contain both supercooled water and ice. This inference is illustrated in Fig. 3.6. The main positive charge involved in lightning flashes probably has a larger vertical extent and is located above the negative charge. An additional, smaller positive charge can be formed below the negative charge.

It appears that the cloud charge structure can be influenced by the updraft speed. Stolzenburg et al. (1998a, b, c) examined data from nearly 50 balloon electric field soundings through convective regions of mesoscale convective systems (MCSs), isolated supercells, and isolated New Mexican mountain thunderclouds. They noted that these three types of thunderclouds may each be characterized by two basic electrical structures, as illustrated in Fig. 3.7. In updrafts of MCS convective regions, in strong updrafts of supercells, and in or near the center of convection in New Mexican thunderclouds, four charge regions were identified. These charge regions can be viewed as the tripolar charge structure described above plus an upper negative

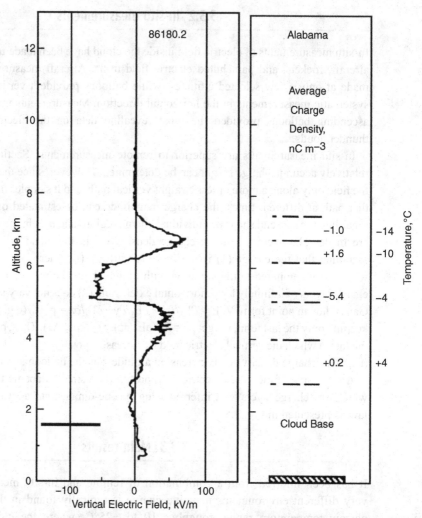

Fig. 3.5 Balloon measurements of the vertical electric field inside a small Alabama thunderstorm. An upward-directed electric field is defined as positive. The values of inferred average charge density (in nC m^{-3}), assuming that charged regions have large horizontal extent and that the field is steady with time, are shown on the right. The field profile is indicative of a "classical" vertical tripole with an upper negative screening layer. Adapted from Marshall and Rust (1991).

screening layer. A more complex charge structure was found to exist outside updrafts of MCS convective regions, outside strong updrafts of supercells, and away from the center of the convection in New Mexican thunderclouds. In these three situations, Stolzenburg et al. (1998a, b, c) identified six or more charge regions, alternating in polarity, with the lowest region being positive.

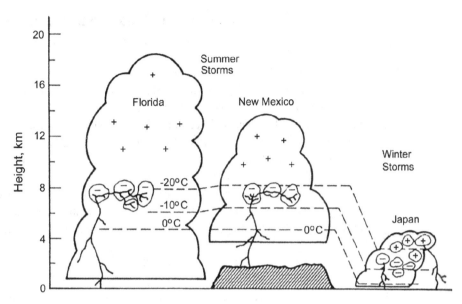

Fig. 3.6 The locations, shown by the small irregular contours inside the cloud boundaries, of ground-flash charge sources observed in summer thunderstorms in Florida and New Mexico and in winter thunderstorms in Japan using simultaneous measurements of electric field at a number of ground stations. Adapted from Krehbiel (1986).

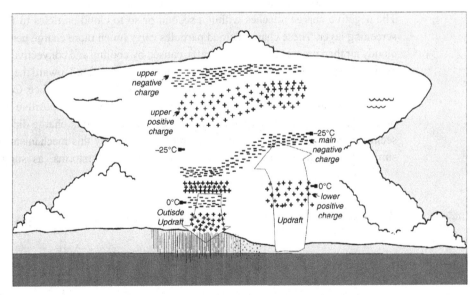

Fig. 3.7 Schematic of the basic charge structure in the convective region of a thunderstorm. Four charge layers are seen near the updraft region, and six charge layers are seen outside the updraft region. The charge structure shown applies to convective elements of mesoscale convective systems (MCS), isolated supercell storms, and New Mexican air-mass storms. Note that there is variability in this basic structure, especially outside the updrafts. Adapted from Stolzenburg et al. (1998b).

3.4 Mechanisms of cloud electrification

Any cloud electrification mechanism involves (1) a small-scale process that electrifies individual hydrometeors and (2) a process that spatially separates these charged hydrometeors by their polarity with the resultant distances between the charged cloud regions being of the order of kilometers. Since most charges reside on hydrometeors of relatively low mobility, the cloud is a relatively good electrical insulator and leakage currents between the charged regions are thought to have a small effect on the charge separation process.

Here, we will only consider the noninductive collisional graupel-ice (or ice-graupel) mechanism and the convection (or convective) mechanism. The term "noninductive" indicates that hydrometeors are not required to be polarized by the ambient electric field. There is a growing consensus that the graupel-ice mechanism is the dominant electrification mechanism.

3.4.1 Convection mechanism

In this mechanism the electric charges are assumed to be supplied by external sources and include: fair-weather and corona space charges near the ground, and charges produced by cosmic rays. Organized convection provides the large-scale separation. According to this mechanism, illustrated in Fig. 3.8, warm air currents (updrafts) carry positive fair-weather space charge to the top of the growing cumulus. Negative charge, produced by cosmic rays above the cloud, is attracted to the cloud's boundary by the positive charges within it. The negative charge attaches within a second or so to cloud particles to form a negative screening layer. These charged cloud particles carry much more charge per unit volume of cloudy air than precipitation. Downdrafts, caused by cooling and convective circulation, are assumed to carry the negative charge down the sides of the cloud toward the cloud base, this negative charge serving to produce positive corona at the Earth's surface. Corona generates additional positive charge under the cloud and, hence, provides a positive feedback to the process. The convective mechanism results in a positive cloud charge dipole, although it seems unlikely that the negative charge region formed by this mechanism would be in a similar temperature range for different types of thunderstorms, as suggested by the

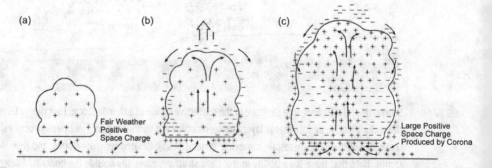

Fig. 3.8   Illustration of the convection mechanism of cloud electrification. Adapted from MacGorman and Rust (1998).

observations. Note that in the convection model there is no role for precipitation in forming the dipole charge structure.

Moore et al. (1989) attempted to demonstrate the feasibility of the convection mechanism by releasing large amounts of negative charge beneath clouds, as opposed to the normally present positive space charge that is postulated to lead to the formation of a positive dipole. If a cloud were to ingest negative space charge and eventually become electrified in the form of a negative dipole (main negative charge above main positive charge), the convection mechanism would be validated. The results of this experiment were inconclusive.

Chauzy and Soula (1999), using a numerical model and measured surface electric fields in Florida and in France, estimated the amount of corona charge that could be transferred from ground to the lower part of cloud by conduction and convection during the lifetime of a thunderstorm. They argue that their estimates of some tens to a few hundred coulombs over an area of 10×10 km^2 for the entire thunderstorm lifetime are comparable to the charge involved in a single lightning flash and, hence, are not in support of the convection mechanism of overall cloud electrification. On the other hand, Chauzy and Soula (1999) suggest that the corona charge transported from the ground surface to the cloud may be responsible for the formation of the lower positive charge region (see Section 3.5).

3.4.2 Graupel-ice collision mechanism

In this mechanism the electric charges are produced by collisions between precipitation particles (graupel) and cloud particles (small ice crystals). Recall that precipitation particles are defined as hydrometeors that have an appreciable fall speed ($\geq$0.3 m s^{-1}), with the hydrometeors that have a lower fall speed, being termed cloud particles. Precipitation particles are generally larger than cloud particles, although there is no absolute demarcation in size to distinguish precipitation particles that are falling out of the cloud from cloud particles which remain essentially suspended or move upward in updrafts. The large-scale separation of charged particles is provided by the action of gravity.

In the graupel-ice mechanism, which appears to be capable of explaining the "classical" tripolar cloud charge structure, the electrification of individual particles involves collisions between graupel particles and ice crystals in the presence of water droplets. The presence of water droplets is necessary for significant charge transfer, as shown by the laboratory experiments. A simplified illustration of this mechanism is given in Fig. 3.9. The heavy graupel particles (two of which are shown in Fig. 3.9) fall through a suspension of smaller ice crystals (hexagons) and supercooled water droplets (dots). The droplets remain in a supercooled liquid state until they contact an ice surface, whereupon they freeze and stick to the surface in a process called riming. Laboratory experiments (e.g. Jayaratne et al., 1983) show that when the temperature is below a critical value called the reversal temperature, T_R, the falling graupel particles acquire a negative charge in collisions with the ice crystals. At temperatures above T_R they acquire a positive charge. The charge sign reversal temperature T_R is generally thought to be between $-10°C$ and $-20°C$, the temperature range characteristic of the main negative charge region found in thunderclouds. The graupel which picks up positive charge when it falls below the altitude of T_R could explain the existence of the

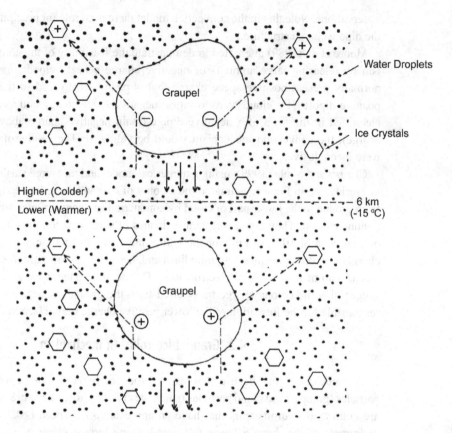

Fig. 3.9 Schematic representation of the graupel-ice mechanism of cloud electrification, in which the charge transfer occurs via collision of graupel with small ice crystals in the presence of supercooled water droplets. It is assumed that the reversal temperature T_R is $-15°C$, and that it occurs at a height of 6 km. © Vladimir A. Rakov and Martin A. Uman 2003, published by Cambridge University Press, reprinted with permission.

lower positive charge region in the cloud, discussed in Section 3.5. Figure 3.10 shows the charge acquired by a simulated riming hail particle during collisions with ice crystals as a function of temperature, from the laboratory experiments of Jayaratne et al. (1983). In general, the sign and magnitude of the electric charge separated during collisions between vapor-grown ice crystals and graupel depends on, besides the temperature, a number of other factors including cloud water content, ice crystal size, relative velocity of the collisions, chemical contaminants in the water, and supercooled droplet size spectrum.

It is believed that the polarity of the charge that is separated in ice-graupel collisions is determined by the rates at which the ice and graupel surfaces are growing. The surface that is growing faster acquires a positive charge. There is no consensus on the detailed physics involved. One reasonable model has been proposed by Baker and Dash (1994). They have suggested that there might be a liquid-like layer (LLL) on ice surfaces and, if so, there

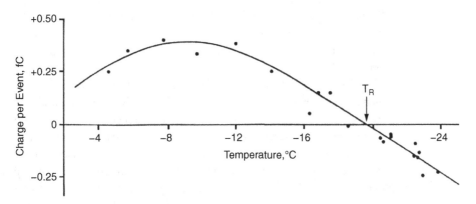

Fig. 3.10 The charge acquired by a riming hail particle (simulated by an ice-covered cylindrical metal rod with diameter 0.5 cm) during collisions with 50-μm ice crystals, as a function of the temperature of the rime in the laboratory. The velocity of impact was 2.9 m s^{-1}. The cloud liquid water content was approximately 1 g m^{-3}, and the mean diameter of water droplets was 10 μm. Charge on the vertical axis is in femtocoulombs (1 fC = 10^{-15} C). Adapted from Jayaratne et al. (1983).

should be an excess negative charge in the outer portion of this layer in the form of OH$^-$ ions. If two ice surfaces collide, the thicker LLL will transfer some of its mass, together with some of the negative ions, to the thinner LLL, and hence leave positive charge behind. Since the surface that is growing faster would tend to have the thickest LLL, this mechanism at least qualitatively describes most of the laboratory data, although more results of recent laboratory experiments of Mason and Dash (2000) show that the theory of Baker and Dash (1994) is only partly correct.

It is possible that the primary electrification mechanism changes once a storm becomes strongly electrified. For example, collisions between ice crystals and graupel could initiate the electrification, and then the larger convective energies of the storm could continue the electrification. It is also possible that important electrification mechanisms are still unrecognized. Recently, Avila et al. (2011) reported new laboratory results on electrification in the temperature range from −37°C and −47°C due to graupel-frozen droplet collisions in the absence of supercooled liquid water. These results may be applicable to the upper parts of clouds (e.g. anvils) where the internal temperatures are substantially lower than −37° C.

3.5 Lower positive charge region

A number of hypotheses have been proposed regarding the origin of the lower positive charge (e.g. Rakov and Uman, Ch. 3). Within the graupel-ice collision mechanism (Section 3.4.2), graupel, which charges positively at temperatures warmer than the reversal temperature, is responsible for the formation of the lower positive charge region. On the other hand, Malan (1952) suggested that the lower positive charge region (LPCR) contains the charge which is produced by corona at the ground and is subsequently carried into the cloud by conduction or convection, and Chauzy and Soula (1999) presented calculations in support of this hypothesis.

While the LPCR may serve to enhance the electric field at the bottom of the negative charge region and thereby facilitate the launching of a negatively charged leader toward ground, the presence of excessive LPCR may prevent the occurrence of negative cloud-to-ground flashes by "blocking" the progression of descending negative leader from reaching ground. Nag and Rakov (2009) inferred four conceptual lightning scenarios that may arise depending upon the magnitude of the LPCR, illustrated in Fig. 3.11. In doing so, they assumed that a pronounced preliminary breakdown (PB) pulse train in electric field records (discussed in Chapter 4) is indicative of interaction of a negatively charged channel extending downward from the main negative charge region with an appreciable LPCR.

When the magnitude of LPCR is abnormally large, say, comparable in magnitude to that of the main negative charge, as shown in Fig. 3.11a (left), the so-called inverted intracloud (IC) discharges are expected to occur. In this scenario, a descending negative leader would likely change its direction of propagation to predominantly horizontal, interact with the LPCR, and be unable to forge its way to ground. The result is an inverted IC flash (attempted cloud-to-ground leader). An example of expected electric field signature of such a discharge is shown in Fig. 3.11a (right), which exhibits a PB pulse train followed by static field change some tens of milliseconds in duration, indicative of an inverted IC flash. If the lightning channel emerges from the cloud, it can be viewed as an "air discharge" or even as a "spider" lightning, if it develops over a large distance near the cloud base.

Figure 3.11b (left) shows the scenario where the magnitude of the LPCR is somewhat smaller than in scenario (a). Similar to scenario (a), a negatively charged leader channel extending vertically from the main negative charge region would become predominantly horizontal, but would eventually make termination on ground. In this case, the discharge can be viewed as a hybrid flash (an IC followed by a cloud-to-ground (CG) discharge). The electric field signature expected for this type of discharge is shown in Fig. 3.11b (right), which shows a PB pulse train followed by a field change characteristic of a cloud discharge lasting for about 50 ms, followed by the first return-stroke waveform of a CG flash.

If the magnitude of the lower positive charge relative to the main negative charge is even smaller, as shown in Fig. 3.11c (left), the descending negative leader would traverse the positive charge region and continue to propagate in a predominantly vertical direction to ground. The electric field signature expected to be produced in this case is shown in Fig. 3.11c (right). It exhibits a PB pulse train and stepped-leader waveform followed by the first return-stroke (RS) waveform. Leader duration, found as the time interval between PB and RS, is about 20 ms.

Figure 3.11d (left) shows the scenario when the LPCR is insignificant. This scenario is similar to scenario (c), except for the LPCR playing essentially no role in negative-leader initiation. The electric field signature produced in this case is expected to be that of a stepped-leader/return-stroke sequence not preceded by a detectable PB pulse train, as shown in Fig. 3.11d (right).

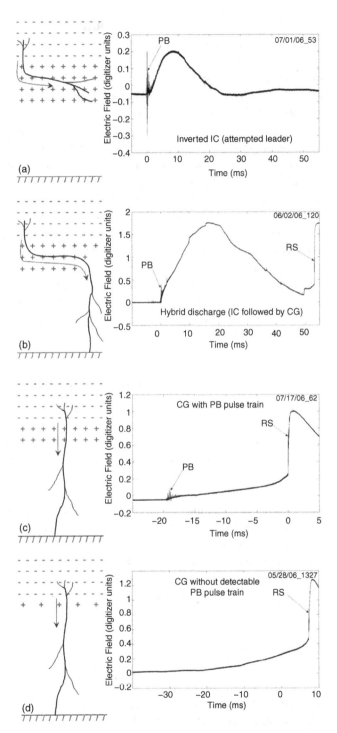

Fig. 3.11 The left panels schematically show four types of lightning that may arise depending upon the magnitude of the LPCR. The charge configuration in each of the scenarios represents only its vertical profile (no lateral boundaries are shown). Arrows indicate the direction of propagation of negative leader. The corresponding examples of expected

3.6 Summary

In calculating cloud fields and lightning-caused field changes at ground level, the typical thundercloud charge structure is often approximated by an idealized model including three vertically stacked point charges (or spherically symmetrical charged volumes): main positive at the top, main negative in the middle, and lower positive at the bottom. This charge configuration is assumed to be located in a non-conducting atmosphere above a perfectly conducting ground. The magnitudes of the main positive and negative charges are typically some tens of coulombs, while the lower positive charge is probably about 10 C or less. The negative charge region tends to have a relatively small vertical extent that is apparently related to the $-10°$ to $-25°$C temperature range, regardless of the stage of storm development, location, and season. Many cloud electrification theories have been proposed. There is growing consensus that the graupel-ice mechanism is the dominant mechanism, at least at the initial stages of cloud electrification. In this mechanism, the electric charges are produced by collisions between graupel and small ice crystals in the presence of water droplets, and the large-scale separation of charged particles is provided by the action of gravity. It is possible that other cloud electrification mechanisms, including the convection mechanism, become important at later stages of the thunderstorm development.

Questions and problems

3.1. What methods can be used to infer the cloud electrical structure? What are their advantages and drawbacks? What is the typical cloud charge structure?

3.2. Discuss the origin of the lower positive charge region and its role in facilitating different types of lightning.

3.3. What is the minimum number of stations needed to infer the charge(s) neutralized by (a) the cloud-to-ground flash and (b) the intracloud flash from multiple-station electric field change measurements?

3.4. What is the polarity of cloud charge region, if the vertical electric field intensity measured at an ascending balloon is (a) positive and increases with increasing height and (b) negative and decreases in magnitude with increasing height?

Caption for Fig. 3.11 (cont.)

electric field signatures are shown in the right panel. The field waveforms are from four different thunderstorms recorded at some tens of kilometers at the Lightning Observatory in Gainesville (LOG), Florida, using the same instrumentation with a decay time constant of 10 ms. PB and RS mark preliminary breakdown pulse train and return-stroke waveform, respectively. Adapted from Nag and Rakov (2009).

3.5. Consider an idealized system of three vertically stacked charges (a vertical tripole) in a non-conducting atmosphere. Assume that the middle negative and top positive charges are 8 and 13 km above ground, respectively, each having a magnitude of 20 C, and that the bottom positive charge is at 2 km and has a magnitude of 2 C. Sketch the electric field intensity at ground due to each of the three charges and due to the entire charge system vs. distance from the axis of the tripole. Estimate the reversal distances.

3.6. For the charge configuration in Problem 5, sketch the electric field change vs. distance, if the top positive and middle negative charges are neutralized via an intracloud discharge.

3.7. Consider a thundercloud whose electrical structure can be represented by two point charges with magnitudes of +40 C and −40 C. The charges are located at a height of 6 km above ground and separated horizontally by 8 km. Sketch the electric field intensity at ground level due to this charge system vs. distance from the point on the ground that is equidistant from the two charges.

3.8. For the charge configuration in Problem 7, sketch the electric field change vs. distance, if the positive charge is neutralized via a cloud-to-ground discharge.

Further reading

Appendix 2: Reconstruction of sources from measured electrostatic field changes

MacGorman, D. R. and Rust, W. D. (1998). *The Electrical Nature of Thunderstorms*, 422 pp., New York: Oxford University Press.

Rakov, V. A. and Uman, M. A. (2003). *Lightning: Physics and Effects*, 687 pp., New York: Cambridge.

Williams, E. R. (1995). Meteorological aspects of thunderstorms. In *Handbook of Atmospheric Electrodynamics*, vol. I, ed. H. Volland, Boca Raton, Florida: CRC Press, 27–60.

Properties of the downward negative lightning discharge to ground

Downward negative lightning discharges – that is, discharges that are initiated in the cloud, initially develop in an overall downward direction, and transport negative charge to ground – probably account for about 90 percent of all cloud-to-ground discharges. The overall cloud-to-ground lightning discharge, termed a "flash," is composed of a number of processes, some of which involve channels that emerge from the cloud, while others involve channels that are confined to the cloud volume. Only processes occurring in channels outside the cloud render themselves to optical observations that can be used to determine channel geometry, extension speed, and other pertinent features of those channels. All lightning processes are associated with the motion of charge and, therefore, can be studied via measurement of the electric and magnetic fields associated with that charge motion.

4.1 General picture

Basic information about downward negative lightning is given in Section 1.2. In the following, referring to Fig. 4.1, we present a more complete sequence of processes involved in a typical downward negative lightning flash. The source of lightning is usually a cumulonimbus (Chapter 3), whose idealized charge structure is shown in Fig. 4.1 as three vertically stacked regions labeled "P" and "LP" for main positive and lower positive charge regions, respectively, and "N" for main negative charge region. The stepped leader is preceded by an in-cloud process called the preliminary or initial breakdown. It may be a discharge bridging the main negative and the lower positive charge regions, as shown in Fig. 4.1 (see also Section 3.5). The initial breakdown serves to provide conditions for the formation of the stepped leader. The latter is a negatively charged plasma channel extending toward the ground at an average speed of 2×10^5 m s^{-1} in a series of discrete steps. From high-speed time-resolved photographs, each step is typically 1 µs in duration and tens of meters in length, with the time interval between steps being 20 to 50 µs. The peak value of the current pulse associated with an individual step has been inferred to be 1 kA or greater. The stepped leader serves to form a conducting path or channel between the cloud charge region and ground. Several coulombs of negative charge are distributed along this path, including downward branches. The leader may be viewed as a process removing negative charge from the cloud charge region and depositing this charge onto the downward extending channel. The stepped-leader duration is typically some tens of milliseconds, and the average leader current is some hundreds of amperes.

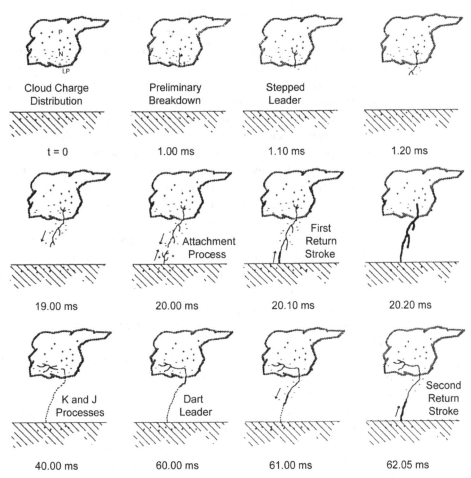

Fig. 4.1 Various processes comprising a two-stroke negative cloud-to-ground lightning flash. Time labels below the sketches can be used to roughly estimate typical durations of the processes and time intervals between them ($t = 0$ corresponds to the beginning of preliminary breakdown process which ends at $t = 1$ ms). Continuing current and M-components are not illustrated in this figure. Adapted from Uman (1987, 2001).

As the leader approaches ground, the electric field at the ground surface, particularly at objects or relief features protruding above the surrounding terrain, increases until it exceeds the critical value for the initiation of one or more upward leaders. It is usually assumed that the initiation of an upward connecting leader (UCL) from ground in response to the descending stepped leader marks the beginning of the attachment process. The attachment process ends when contact is made between the hot channels of the downward and upward moving leaders, probably some tens of meters above ground (more above a tall structure), where the first return stroke begins. The return stroke serves to neutralize the leader charge or, equivalently, to transport the negative charges stored on the leader channel to the ground.

It is worth noting that the return-stroke process may not neutralize all the leader charge and is likely to deposit some excess positive charge onto the upper part of the leader channel and into the cloud charge source region. The final stage of the attachment process and the initial stage of the return-stroke process are complex and will be described in Sections 4.4 and 4.5, respectively. The net result of those stages is a fully formed return stroke, similar to the potential discontinuity that would travel upward along a vertical, negatively charged transmission line if the lower end of the line were connected to the ground. The speed of the return stroke, averaged over the visible channel, is typically between one-third and one-half of the speed of light. There is no consensus about whether or how the first return-stroke speed changes over the lower 100 m or so, but over the entire channel, the speed decreases with increasing height, dropping abruptly after passing each major branch. At the same time, a transient enhancement of the channel luminosity below the branch point, termed a branch component, is often observed.

The first return-stroke current measured at the ground rises to an initial peak of about 30 kA in some microseconds and decays to half-peak value in some tens of microseconds while exhibiting a number of subsidiary peaks, probably associated with branches. This impulsive component of current may be followed by a current of some hundreds of amperes lasting for some milliseconds. The return stroke effectively lowers to the ground the several coulombs of charge originally deposited on the stepped-leader channel including all the branches.

When the first return stroke, including any associated in-cloud discharge activity (discussed later), ceases the flash may end. In this case, the lightning is called a single-stroke flash. However, more often the residual first-stroke channel is traversed by a leader that appears to move continuously, termed a dart leader. During the time interval between the end of the first return stroke and the initiation of a dart leader, J (for junction) and K processes occur in the cloud. K processes can be viewed as transients occurring during the slower J process. The J processes amount to a redistribution of cloud charge on a tens-of-milliseconds timescale in response to the preceding return stroke. There is controversy as to whether these processes, which apparently act to extend the return-stroke channel further into the cloud, are necessarily related to the initiation of a following dart leader. The J process is often viewed as a relatively slow positive leader extending from the flash origin into the negative charge region, with the K process being a relatively fast "recoil process" that begins at the tip of the positive leader or in a decayed positive leader branch and propagates toward the flash origin. Both the J processes and the K processes in cloud-to-ground discharges serve to transport additional negative charge into and along the existing channel (or its remnants), although not all the way to the ground. In this respect, K processes may be viewed as attempted dart leaders. The processes that occur after the only stroke in single-stroke flashes and after the last stroke in multiple-stroke flashes are sometimes termed F (for final) processes. These are similar, if not identical, to J processes.

The dart leader progresses downward at a typical speed of $10^7\ \mathrm{m\ s^{-1}}$, typically ignores the first stroke branches, and deposits along the channel a total charge of the order of 1 C. The dart leader current peak is about 1 kA. Some leaders exhibit stepping near ground while propagating along the path traversed by the preceding return stroke; these leaders are termed "dart-stepped leaders." Additionally, some dart or dart-stepped leaders deflect from the previous return-stroke path, become stepped leaders, and form a new termination on the ground.

When a dart leader or dart-stepped leader approaches the ground, an attachment process similar to that described for the first stroke takes place, although it probably occurs over a shorter distance and consequently takes less time, with the upward connecting leader length being of the order of some meters. Once the bottom of the dart or dart-stepped leader channel is connected to the ground, the second (or any subsequent) return-stroke wave is launched upward, which again serves to neutralize the leader charge. The subsequent return-stroke current at ground typically rises to a peak value of 10 to 15 kA in less than a microsecond and decays to half-peak value in a few tens of microseconds. The upward propagation speed of subsequent return strokes is similar to that of first return strokes, although, due to the absence of branches, the speed variation along the channel does not exhibit abrupt drops.

The impulsive component of the current in a subsequent return stroke is often followed by a continuing current which has a magnitude of tens to hundreds of amperes and a duration up to hundreds of milliseconds. Continuing currents with a duration in excess of 40 ms are traditionally termed "long continuing currents." Between 30 percent and 50 percent of all negative cloud-to-ground flashes contain long continuing currents. The source for continuing current is the cloud charge, as opposed to the charge distributed along the leader channel, the latter charge contributing to at least the initial few hundred microseconds of the return-stroke current observed at ground. Continuing current typically exhibits a number of superimposed surges that rise to peak in tens to hundreds of microseconds, with the peak being generally in the hundreds of amperes range but occasionally in the kiloamperes range. These current surges are associated with enhancements in the relatively faint luminosity of the continuing-current channel. Both the current surges and luminosity enhancements are called M components. Note that continuing current and M-component processes are not shown in Fig. 4.1.

The time interval between successive return strokes in a flash is usually several tens of milliseconds (see Fig. 4.2), although it can be as large as many hundreds of milliseconds if a long continuing current is involved, or as small as one millisecond or less. Note that inter-stroke intervals are usually measured between the peaks of current or electromagnetic field pulses. The total duration of a flash is typically some hundreds of milliseconds, and the total charge lowered to ground is some tens of coulombs. The average numbers of strokes per flash and percentages of single-stroke flashes observed in different locations are summarized in Table 4.1. It follows from Table 4.1 that the overwhelming majority (about 80 percent or more in most cases) of negative cloud-to-ground flashes contain more than one stroke.

One-third to one-half of all lightning discharges to earth, both single- and multiple-stroke flashes, strike the ground at more than one point (see Table 4.2) with the spatial separation between the channel terminations being up to many kilometers (see Fig. 4.3). According to Table 4.2, the average number of channels per flash is remarkably similar for different geographical locations, mostly ranging from 1.5 to 1.7. In most cases, multiple ground

Table 4.1 Number of strokes per negative flash and percentage of single-stroke flashes. Adapted from CIGRE TB 549 (2013)

Location (Reference)	Average Number of Strokes per Flash	Percentage of Single-Stroke Flashes	Sample Size
New Mexico (Kitagawa et al., 1962)	6.4	13%	83
Florida (Rakov and Uman, 1990b)	4.6	17%	76
Sweden (Cooray and Perez, 1994)	3.4	18%	137
Sri Lanka (Cooray and Jayaratne, 1994)	4.5	21%	81
China (Qie et al., 2002)	3.8	40%	83
Arizona (Saraiva et al., 2010)	3.9	19%	209
Brazil (Ballarotti et al., 2012)	4.6	17%	883
Malaysia (Baharudin et al., 2014)	4.0	16%	100

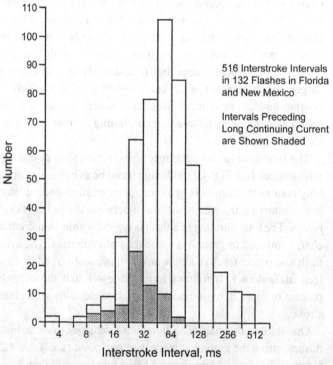

516 Interstroke Intervals in 132 Flashes in Florida and New Mexico

Intervals Preceding Long Continuing Current are Shown Shaded

Fig. 4.2 Histogram of 516 inter-stroke intervals in 132 flashes in Florida and New Mexico. Intervals preceding strokes that initiated long continuing currents (Section 4.7) are shown shaded. Adapted from Rakov and Uman (1990a).

Table 4.2 Number of channel terminations per flash and percentage of multi-grounded flashes. Adapted from CIGRE TB 549 (2013)

Location (Reference)	Average Number of Channels per Flash	Percentage of Multi-Grounded Flashes	Sample Size
New Mexico	1.7	49%	72[*]
(Kitagawa et al., 1962)	1.6	42%	83[**]
Florida	1.7	50%	76
(Rakov and Uman, 1990b)			
France	1.5	34%	2995
(Berger et al., 1996; Hermant, 2000)			
Arizona	1.4	35%	386
(Valine and Krider, 2002)			
US Central Great Plains	1.6	33%	103
Fleenor et al. (2009)			
Brazil	1.7	51%	138
(Saraiva et al., 2010)			
Arizona	1.7	48%	206
(Saraiva et al., 2010)			

[*] multiple-stroke flashes only
[**] including 11 single-stroke flashes, each assumed to have a single channel per flash

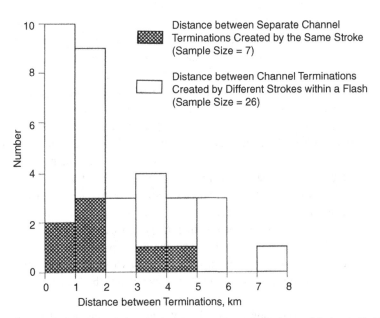

Fig. 4.3 Histogram of the distances between the multiple terminations of 22 individual ground flashes in Florida. Adapted from Thottappillil et al. (1992).

terminations within a given flash are associated not with an individual multi-grounded leader but rather with the deflection of a subsequent leader from the previously formed channel.

Examples of simultaneous photographic and electric field records of two negative multiple-stroke flashes that additionally illustrate the various lightning processes described above are given in Fig. 4.4. The salient properties of downward negative lightning discharges are summarized in Table 4.3.

4.2 Initial breakdown

4.2.1 General information

The initial breakdown, often referred to as preliminary breakdown, of a cloud-to-ground flash is the in-cloud process that initiates or leads to the initiation of the downward-moving stepped leader. In the early lightning studies in South Africa, the existence of the initial breakdown as a unique lightning process was generally inferred from (1) observations of luminosity produced by thunderclouds for a hundred or more milliseconds before the emergence of the stepped leader from the cloud base, (2) observations of relatively long electric field changes, exceeding 100 ms in duration, prior to the first return stroke, and (3) the assumption that the stepped-leader duration was unlikely to exceed a few tens of milliseconds based on the measured leader speed below the cloud and the cloud charge height. Clarence and Malan (1957) suggested, on the basis of single-station electric field measurements of the type illustrated in Fig. 4.5, that the initial breakdown (labeled B by them) is a vertical discharge between the main negative charge region and the lower positive charge region that has a duration of 2 to 10 ms. This inference is made from the observed polarity reversal of the B portion of the electric field waveforms of the type shown in Fig. 4.5 in the distance range from 2 to 5 km. According to Clarence and Malan (1957), the initial breakdown is followed by the stepped leader (labeled L), either immediately or after the so-called intermediate stage (labeled I) that may last up to 400 ms. The intermediate stage was interpreted by Clarence and Malan (1957) as being due to negative charging of the vertical channel of the initial breakdown until the field at the bottom of the channel was high enough to launch a stepped leader which initiated, on its arrival at ground, a return stroke labeled R in Fig. 4.5. Clarence and Malan's (1957) scenario of the initial breakdown, inferred from single-station electric field measurements, is not necessarily confirmed by more recent studies based on multiple-station electric field measurements or on VHF channel imaging in conjunction with electric field records. These more recent studies suggest that the initial breakdown can often be viewed as a sequence of channels extending in seemingly random directions from the cloud charge source. One of these events evolves into the stepped leader which bridges the cloud charge source and the ground.

Figure 4.6 shows 274 MHz interferometric images of the initial breakdown in a six-stroke Florida flash (Shao, 1993). The process develops in three successive branches

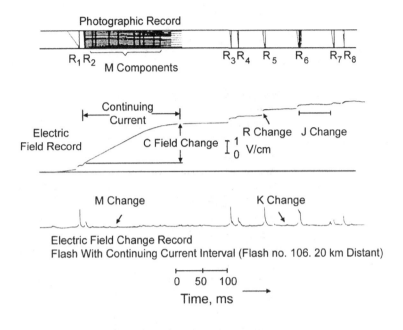

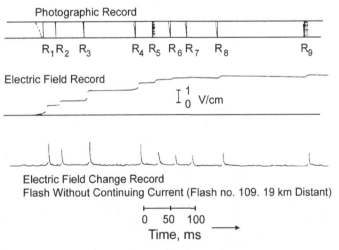

Photographic Record

R₁ R₂ **M Components** R₃ R₄ R₅ R₆ R₇ R₈

Electric Field Record

Continuing Current

C Field Change R Change J Change

M Change K Change

Electric Field Change Record
Flash With Continuing Current Interval (Flash no. 106. 20 km Distant)

0   50 100
Time, ms

Photographic Record

R₁ R₂ R₃ R₄ R₅ R₆ R₇ R₈ R₉

Electric Field Record

Electric Field Change Record
Flash Without Continuing Current (Flash no. 109. 19 km Distant)

0   50 100
Time, ms

Fig. 4.4 Simultaneous photographic and electric field measurements for two multiple-stroke ground flashes in New Mexico. The upper three diagrams relate to a flash with long continuing current (20 km distant) and the lower three diagrams to a flash without long continuing current (19 km distant). Two electric field records are shown for each flash. One, labeled "Electric Field Record," was obtained with a measuring system having a decay time constant of 4 s and a relatively low gain. This and other systems with decay time constants of the order of seconds are sometimes referred to as "slow antenna" systems. The other electric field record, labeled "Electric Field Change Record," was obtained with a measuring system having a decay time constant of 70 μs and a relatively high gain. This and other systems having a sub-millisecond decay time constant are sometimes referred to as "fast antenna" systems. Adapted from Kitagawa et al. (1962).

Table 4.3 Characterization of negative cloud-to-ground lightning. Adapted from Rakov and Uman (2003)

Parameter	Typical Value[a]
Stepped leader	
Step length, m	50
Time interval between steps, μs	20–50
Step current, kA	>1
Step charge, mC	>1
Average propagation speed, m s^{-1}	2×10^5
Overall duration, ms	35
Average current, A	100–200
Total charge, C	5
Electric potential, MV	~ 50
Channel temperature, K	~ 10,000
First return stroke[b]	
Peak current, kA	30
Maximum current rate of rise, kA μs^{-1}	10–20
Current rise time (10–90 percent), μs	5
Current duration to half-peak value, μs	70–80
Charge transfer, C	5
Propagation speed, m s^{-1}	$(1–2) \times 10^8$
Channel radius, cm	~ 1–2
Channel temperature (peak), K	~ 30,000
Dart leader	
Speed, m s^{-1}	$(1–2) \times 10^7$
Duration, ms	1–2
Charge, C	1
Current, kA	1
Electric potential, MV	~ 15
Channel temperature, K	~ 20,000
Dart-stepped leader	
Step length, m	10
Time interval between steps, μs	5–10
Average propagation speed,	$(1–2) \times 10^6$ m s^{-1}
Subsequent return stroke[b]	
Peak current, kA	10–15
Maximum current rate of rise, kA μs^{-1}	100
10–90 percent current rate of rise, kA μs^{-1}	30–50
Current rise time (10–90 percent), μs	0.3–0.6
Current duration to half-peak value, μs	30–40
Charge transfer, C	1
Propagation speed, m s^{-1}	$(1–2) \times 10^8$
Channel radius, cm	~ 1–2
Channel temperature (peak), K	~ 30,000

Table 4.3 (cont.)	
Parameter	Typical Value
Continuing current (longer than 40 ms or so)[c]	
Magnitude, A	100–200
Duration, ms	~ 100
Charge transfer, C	10–20
M component[b]	
Peak current, A	100–200
Current rise time (10–90 percent), μs	>100
Charge transfer, C	0.1–0.2
Overall flash	
Duration, ms	200–300
Number of strokes per flash[d]	3–5
Inter-stroke interval, ms	60
Charge transfer, C	20
Energy, J	10^9–10^{10}

[a] Typical values are based on a comprehensive literature search and unpublished experimental data acquired by the University of Florida Lightning Research Group.
[b] All current characteristics for return strokes and M components are based on measurements at the lightning channel base.
[c] About 30 to 50 percent of lightning flashes contain continuing currents longer than 40 ms or so.
[d] About 15 to 20 percent of lightning flashes are composed of a single stroke.

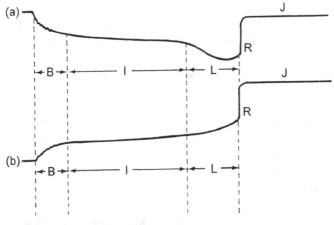

Fig. 4.5 Examples of electric field waveforms of the type used by Clarence and Malan (1957) to introduce the so-called BIL structure of the electric field prior to the first return stroke and to interpret the B stage as a vertical discharge between the main negative charge region and the lower positive charge region in the cloud: (a) the electric field at 2 km, (b) the electric field at > 5 km. Adapted from Clarence and Malan (1957).

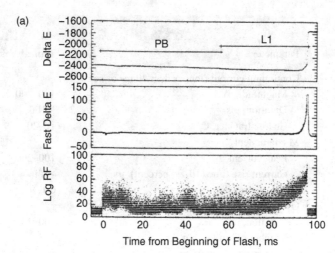

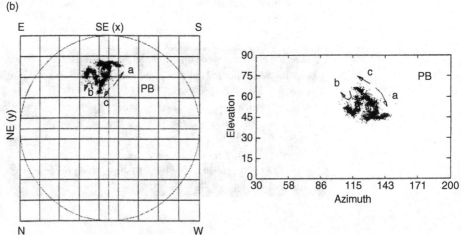

Fig. 4.6   Preliminary breakdown inside the cloud: (a) time waveforms from the beginning of the flash to the first return stroke with the initial (preliminary) breakdown indicated by PB and leader indicated by L1, (b) the radiation sources of PB. The atmospheric electricity sign convention, according to which a downward-directed electric field or field change is defined as positive, is used in (a). Adapted from Shao (1993).

originating from the volume that appears to also be the source for the following stepped leader. The latter is not shown in Fig. 4.6. The duration of the entire process is 50 to 60 ms.

It is worth noting that considerable intra-cloud discharge activity can precede ground flashes, apparently not related directly to the initiation of stepped leaders. Furthermore, strokes to the ground sometimes appear to be a minor branch of an extensive cloud flash.

4.2.2 Initial breakdown pulses

It has been observed by many investigators that the first return stroke electric field waveform (Section 4.5) may be preceded by a train of relatively large microsecond-scale bipolar

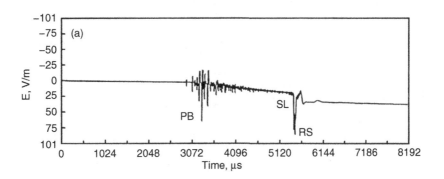

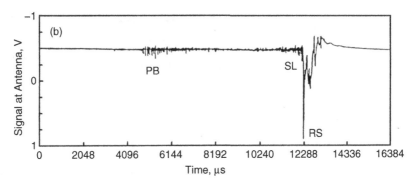

Fig. 4.7 Examples of electric fields due to negative first strokes in cloud-to-ground lightning: (a) winter lightning at about 25 km, (b) summer lightning at unknown distance. PB stands for preliminary breakdown, SL for stepped leader (just prior to its attachment to ground), and RS for return stroke. Overall timescales in Figs. 4.7a and b are about 8.2 and 16.4 ms, respectively. Note that the separation between PB and RS, the duration of the stepped leader, is smaller than usual, particularly in Fig. 4.7a. The atmospheric electricity sign convention, according to which a downward directed-electric field or field change is defined as positive, is used here. Adapted from Brook (1992).

pulses, as illustrated in Figs. 4.7a and b. The duration of the train is a few milliseconds or less. Nag and Rakov (2009) suggested that a pronounced preliminary breakdown (PB) pulse train in electric field records is indicative of interaction of a negatively charged channel extending downward from the main negative charge region with an appreciable lower positive charge region (LPCR) (see Section 3.5). In this view, the PB is just the initial stage of stepped leader when it moves through the LPCR. The pulses in the train are usually bipolar with the initial polarity being the same as that of the following return-stroke pulse. The amplitude of the initial breakdown pulses can be comparable to that of the first return stroke, as seen in Fig. 4.7a, and sometimes even exceed it. On the other hand, in some records the initial breakdown pulses are either undetectable or have negligible amplitude compared to the following return stroke pulses. Some investigators considered the occurrence of pronounced PB pulses followed by a period of many milliseconds of relatively small, if any, pulses as indicative of a β-type leader (Section 4.3). Brook (1992) found that the initial breakdown pulses have larger amplitudes in winter negative lightning than in

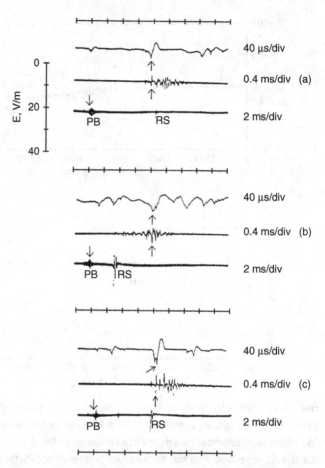

Fig. 4.8 Electric fields radiated by cloud-to-ground lightning discharges at distances of 50–100 km. The arrows indicate the same event on timescales of 2 ms (bottom traces), 0.4 ms (middle traces), and 40 μs (top traces) per division. The atmospheric electricity sign convention, according to which a downward-directed electric field or field change is defined as positive, is used here. PB stands for preliminary breakdown and RS for return stroke. Adapted from Weidman and Krider (1979).

summer negative lightning (see Figs. 4.7a, b) and attributed the disparity to the difference in the precipitation mix in summer and winter clouds. However, other investigators observed pronounced PB pulse trains under a variety of conditions, including summer thunderstorms. The occurrence of pronounced PB pulses appears to depend on the type of storm and the stage of storm life cycle.

Examples of initial breakdown pulses as they appear in electric field records taken 50–100 km from the lightning discharge are given in Fig. 4.8. Individual pulses are characterized by a total duration of 20–40 μs. The average number of large pulses per train is about 10. The pulse rise time is typically 10 μs. Usually there are two or three smaller pulses superimposed on the rising portion of the initial breakdown pulse, while the falling portion and the opposite polarity overshoot are smooth. It is worth noting that the initial breakdown pulses are distinctly different from the stepped-leader pulses produced near ground (see Fig. 4.22). The differences are seen in terms of (1) the overall waveshape: the former are bipolar with fine structure and the latter are essentially unipolar and smooth, (2) the total duration, 20–40 µs vs. 1–2 µs, and (3) the inter-pulse interval, 70–130 µs vs. 15–25 µs. The characteristics, including the polarity of the initial half cycle, of the initial breakdown pulses in negative ground flashes differ from those in cloud flashes.

4.3 Stepped leader

4.3.1 General information

A streak photograph of a negative downward stepped leader is shown in Fig. 4.9. It is worth noting that in the studies which have resulted in the bulk of the available information on stepped leaders, leader steps were difficult to photograph and could be reproduced well only by intensifying the negative. In early lightning studies (e.g. Schonland, 1938; Schonland et al., 1938a, b), the variety of observed stepped-leader forms was organized in two large categories: α-type leaders and β-type leaders. The majority of photographed leaders were of the α-type. These are characterized by a uniform downward speed of the order of 10^5 m s^{-1} and steps that do not vary appreciably in length or brightness and are shorter and much less luminous than for β-type leaders. In contrast with α-leaders, β-leaders appear to involve two stages in their development outside the cloud. They begin beneath the cloud base or

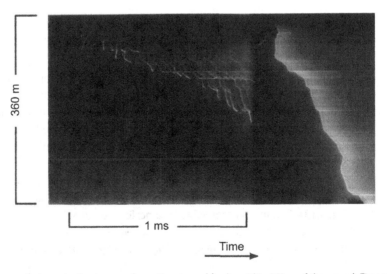

Fig. 4.9 Streak-camera photograph of a downward negative stepped leader within 360 m of the ground. Time advances from left to right. The left part of the photograph was overexposed in the reproduction process in order to enhance the intensity of the early portion of the leader image. Adapted from Berger and Vogelsanger (1966).

occasionally emerge from the side of the cloud with relatively long, bright steps and high average speed, of the order of 10^6 m s^{-1}, and exhibit extensive branching near the cloud base. As a β-leader approaches the ground, it assumes the characteristics of α-leaders: that is, it exhibits lower propagation speed and shorter and less luminous steps. Both α and β leaders usually exhibit an increased average speed and increased step brightness as they approach the ground. Schonland et al. (1938b) further consider two subgroups, labeled $β_1$ and $β_2$, within the β-type leaders. $β_1$-leaders are defined as having an abrupt discontinuity in downward speed at some point in their trip from the cloud to the ground. The second stage occurs within the bottom kilometer or so of the channel and is similar to a "normal" α-leader. It appears that the $β_1$-leader is the same as the previously defined β-leader. The $β_2$-leader is similar to the $β_1$-leader, except that during the second, α-leader stage the channel is traversed by one or more luminosity waves continuously propagating from the cloud to the stepped-leader tip. These waves are separated by time intervals of the order of 10 ms, and each one resembles the dart leader discussed in Section 4.6. As a much faster moving and much more luminous "dart" catches up with the downward-moving bottom end (tip) of the stepped leader, profuse outward branching momentarily occurs at the tip. $β_2$-leaders are probably rare. It is quite possible that stepped leaders classified as α-leaders actually are similar to β-leaders whose more branched, faster, and brighter upper part is hidden inside the cloud.

The stepped leader propagates toward the ground at an average speed of 2×10^5 m s^{-1}. The average stepped-leader current is some hundreds of amperes. The typical duration of the stepped leader estimated from measured electric field waveforms is some tens of milliseconds and the total charge is several coulombs. Rakov and Uman (1990c) reported distributions of leader duration for strokes of different order in Florida, reproduced in Fig. 4.10. They found a geometric mean leader duration of 35 ms for stepped leaders.

4.3.2 Overall electric and magnetic fields

Examples of the electric field changes produced by stepped leaders together with the corresponding return-stroke field changes as a function of time are given in Figs. 4.11 and 4.12 (left panel, the top Delta E waveform). We will discuss next, with reference to Fig. 4.13, equations that can be used to describe the stepped-leader electric and magnetic fields. These field equations are also applicable to the subsequent leaders discussed in Section 4.6.

The leader is assumed to create a channel extending vertically downward with a constant speed v from a stationary and spherically symmetrical charge source at height H_m (see Fig. 4.13). Ground is assumed to be a perfect conductor.

The total leader electric field is comprised of electrostatic, induction, and radiation field components and the total magnetic field of magnetostatic and radiation field components. Equations for the total electric and magnetic fields, including their detailed derivations, are found in Thotttappillil et al. (1997; Eqs. B38 and B39, respectively). In the following, we will consider only the electrostatic approximation for the leader electric field and the magnetostatic approximation for the leader magnetic field that are valid when the

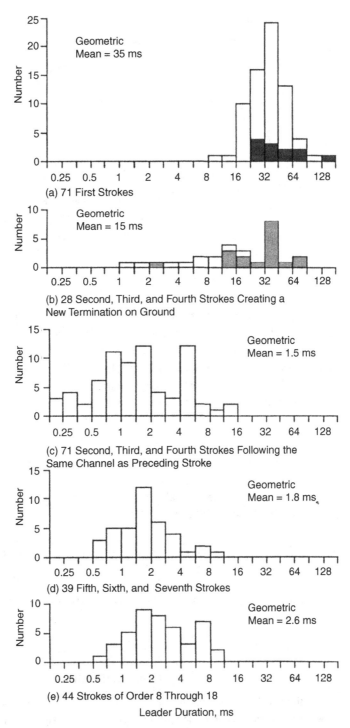

(a) 71 First Strokes

(b) 28 Second, Third, and Fourth Strokes Creating a New Termination on Ground

(c) 71 Second, Third, and Fourth Strokes Following the Same Channel as Preceding Stroke

(d) 39 Fifth, Sixth, and Seventh Strokes

(e) 44 Strokes of Order 8 Through 18

Leader Duration, ms

Fig. 4.10 Histograms of leader duration for five sets of strokes: (a) 71 first strokes; (b) 28 second, third, and fourth strokes creating a new termination on ground; (c) 71 second, third, and fourth strokes following the same channel as the preceding stroke; (d) 39 fifth, sixth, and seventh strokes, and (e) 44 strokes of order 8 through 18.

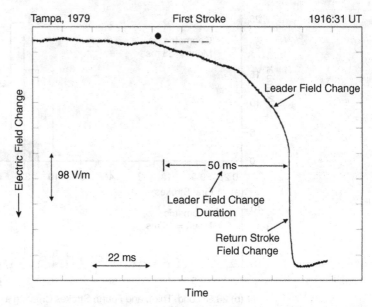

Tampa, 1979   First Stroke   1916:31 UT

Electric Field Change

Leader Field Change

98 V/m

50 ms

Leader Field Change
Duration

Return Stroke
Field Change

22 ms

Time

Fig. 4.11 Electric field change for the first stroke in a three-stroke flash that occurred in Florida at a distance of about 12 km. A small solid circle marks the starting point of the leader field change. The time interval labeled 50 ms shows the leader field change duration. Electric field changes due to the leader and due to the return stoke are labeled. A positive field change deflects downward. The atmospheric electricity sign convention, according to which a downward-directed electric field or field change is defined as positive, is used here. The leader field change is monotonic positive. Adapted from Rakov and Uman (1990c).

significant wavelengths of the electric and magnetic fields are much larger than the dimensions of the overall system of the lightning and the observer. These approximations are expected to be applicable to overall fields produced by close lightning leaders.

We will start with the general expression for the electrostatic component of the leader field given by Thottappillil et al. (1997),

$$E_z(r,\ t) = \frac{1}{2\pi\varepsilon_0} \int_{h(t)}^{H_m} \frac{z'}{R^3(z')} \rho_L \left(z',\ t - \frac{R(z')}{c} \right) dz'$$
$$- \frac{1}{2\pi\varepsilon_0} \frac{H_m}{R^3(H_m)} \int_{h(t)}^{H_m} \rho_L \left(z',\ t - \frac{R(z')}{c} \right) dz' \tag{4.1}$$

Caption for Fig. 4.10 (cont.)

Shown shaded in Fig. 4.10a are the data for single-stroke flashes, while in Fig. 4.10b the shading indicates a new termination on ground involving a path between cloud base and ground that is completely separate from that of the preceding stroke as opposed to having a common portion of the channel seen beneath the cloud. Adapted from Rakov and Uman (1990c).

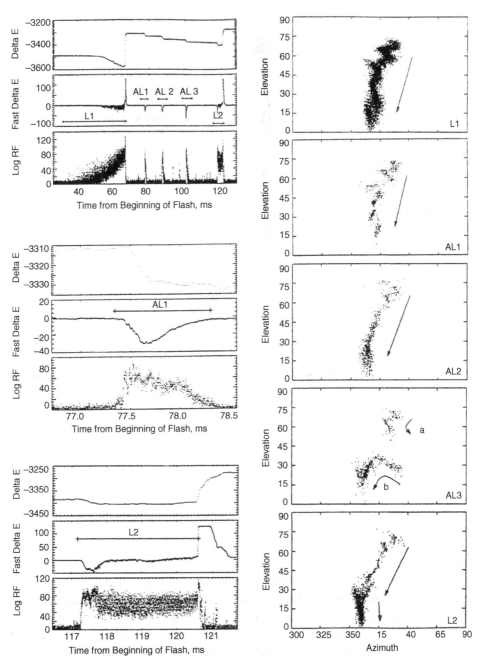

Fig. 4.12   Electric field (Delta E and Fast Delta E), logarithmic RF radiation amplitude (log RF), and radiation sources mapped in the elevation-azimuth coordinates for (1) a stepped leader (L1), (2) three "attempted" leaders (AL1, AL2, and AL3), and (3) a subsequent leader creating a new termination on the ground (L2), that occurred during the first 120 ms of a multiple-stroke cloud-to-ground Florida flash. The atmospheric electricity sign convention, according to which a downward-directed electric field or field change is defined as positive, is used here. The stepped-leader field change is hook-shaped net negative. Adapted from Shao et al. (1995).

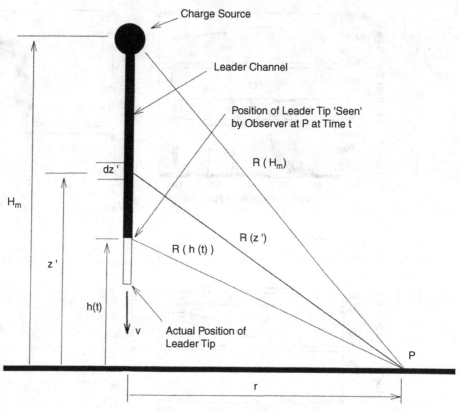

Fig. 4.13 The geometry used in deriving expressions for electric and magnetic fields at a point P on Earth a horizontal distance r from the vertical lightning leader channel extending downward with speed *v*. Adapted from Thottappillil et al. (1997).

where $R(H_m) = (H_m^2 + r^2)^{1/2}$ (see Fig. 4.13) and $h(t)$ is a height at which the observer "sees" the lower end of the leader channel, given by the solution of

$$t = \frac{H_m - h(t)}{v} + \frac{\sqrt{h^2(t) + r^2}}{c} \tag{4.2}$$

The first term of Eq. 4.1 represents the field change due to the charge on the downward-extending leader channel, and the second term represents the field change due to the depletion of the charge at the cloud charge source as it is "drained" by the extending leader channel. The total charge on the leader channel at any time is equal to the total charge removed from the cloud charge source, so that the net charge on the overall channel/source system is zero at all times. The placement of opposite-polarity charge at the origin simulates the more or less spherically symmetrical development of a large number of branched channels whose function is to collect negative charges from the hydrometeors and "funnel" them to the downward-progressing channel. One can view the entire leader system as being composed of negative and positive sections, the negative section extending vertically toward the ground and the positive section being heavily branched inside the cloud.

We now assume that the maximum difference in propagation time from any source on the channel to the observer is much less than the time required for significant variation in the sources (i.e. retardation effects are negligible), and rewrite Eq. 4.1 as

$$E_z(r, \; t) = \frac{-1}{2\pi\varepsilon_0} \int_{H_{\rm m}}^{z_t} \left[\frac{z'}{R^3(z')} - \frac{H_{\rm m}}{R^3(H_{\rm m})} \right] \rho_{\rm L}(z', \; t) \; dz' \tag{4.3}$$

where $z_t = H_{\rm m} - vt$ is the height of the leader tip at time t and v is the leader speed, assumed to be a constant.

If we assume that $\rho_{\rm L}(z', \; t) = \rho_{\rm L} = $ constant, which corresponds to a uniformly charged leader channel, we can rewrite Eq. 4.3 in the following form

$$E_z(r, \; t) = \frac{\rho_{\rm L}}{2\pi\varepsilon_0 r} \left[\frac{1}{(1 + z_t^2/r^2)^{1/2}} - \frac{1}{(1 + H_{\rm m}^2/r^2)^{1/2}} - \frac{(H_{\rm m} - z_t)H_{\rm m}}{r^2(1 + H_{\rm m}^2/r^2)^{3/2}} \right] \tag{4.4}$$

where the first two terms represent the contribution from the leader channel and the third term the contribution from the source. The leader electric field changes computed from Eq. 4.4 are shown in Fig. 4.14 as a function of normalized time t/T, where $T = H_{\rm m}/v$, for different normalized distances $r/H_{\rm m}$ from the channel. If, for example, $H_{\rm m} = 5$ km, the distance range represented in Fig. 4.14 is from 3.5 to 7 km. Leader field changes calculated by Rubinstein et al. (1995) at closer distances, $r = 30$ m and $r = 500$ m, are shown in Figs. 4.15a and b, respectively. Note that the latter field changes were computed for different values of speed, all characteristic of subsequent leaders. The shape of leader electric field change waveform is determined by relative contributions from the channel (negative) and from the source (positive). This is why, as the distance increases in Fig. 4.14, the leader field change waveshape transforms from hook-shaped net negative to hook-shaped net positive, and then to monotonic positive. The net leader field change, which in Fig. 4.14 corresponds to $t/T = 1.0$, is equal to zero at $r/H_{\rm m} = 0.79$ (tan $\alpha_{\rm s} = H_{\rm m}/r = 1.27$, where $\alpha_{\rm s} = 52°$ is the angle between the line joining the source and the observation point and the ground surface). [Note that the curve labeled 0.8 in Fig. 4.14 actually corresponds to $r/H_{\rm m} = 0.77$, rounded off to 0.8.] In Fig. 4.15, the distances are very small (the contribution from the channel is dominant) and hence all the leader field changes are negative.

For a very close field point where $H_{\rm m} \gg 2 \, r$, and for either $z_t = 0$ (leader touching the ground) or $z_t^2 \ll r^2$ and $z_t \ll H_{\rm m}$ (leader close to the ground), Eq. 4.4 becomes approximately (Rubinstein et al., 1995)

$$E_z(z_t = 0) \approx \frac{\rho_{\rm L}}{2\pi\varepsilon_0 r} \tag{4.5}$$

That is, very close to the channel, the vertical electrostatic field change at ground due to the fully developed leader channel falls off with distance as r^{-1}, as opposed to r^{-3} variation far from the channel ($H_{\rm m}^2 \ll r^2$). Interestingly, Eq. 4.5 is exactly the same expression as the radial field that would be produced by an infinitely long, uniform line charge in free space.

The magnetostatic approximation for the leader magnetic field is usually written in terms of current $I(t)$ which is assumed to be slowly varying and the same at all heights along the vertical lightning channel (e.g. Uman, 1987, 2001). A less familiar formulation in terms

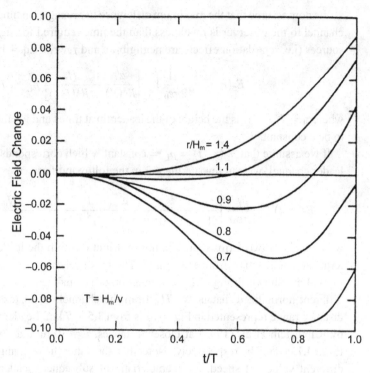

Fig. 4.14   Waveshapes of negative leader electric field changes at different distances r. H_m is the height of the cloud charge source, t is the time, and $T = H_m/v$ is the time required for a leader propagating from the source at H_m at a constant speed v to reach ground. If $H_m = 5$ km, the distance range represented in this figure is from 3.5 to 7 km. The net leader field change corresponds to $t/T = 1.0$ and is equal to zero at $r/H_m = 0.79$ ($H_m/r = 1.27$). In order to obtain V m^{-1}, multiply the ordinate by $|\rho_L| (2\pi\varepsilon_0 r)^{-1}$, where the magnitude of line charge density $|\rho_L|$ is in C m^{-1}, $\varepsilon_0 = 8.85 \times 10^{-12}$ F m^{-1}, and r is in meters. The atmospheric electricity sign convention, according to which a downward-directed electric field or field change is defined as positive, is used here. © Vladimir A. Rakov and Martin A. Uman 2003, published by Cambridge University Press, reprinted with permission.

of charge density ρ_L for the leader magnetic field is given by Thottappillil et al. (1997). The magnetostatic field of a vertical leader channel, the top end of which is at height H_m and the bottom end at height $z_t = h(t)$ (Fig. 4.13), is given by

$$B_\phi(r,\, t) = \frac{\mu_0}{2\pi r}\left[\frac{H_m}{R(H_m)} - \frac{z_t}{R(z_t)}\right] I(t) \tag{4.6}$$

For a fully developed leader channel, that is, for $z_t = 0$,

$$B_\phi(r,\, t) = \frac{\mu_0}{2\pi r}\left[\frac{H_m}{R(H_m)}\right] I(t) \tag{4.7}$$

Equation 4.7 is an expression for the magnetostatic field of a vertical current-carrying line the bottom end of which is at ground (a perfect conductor) and the top end is at height H_m. If

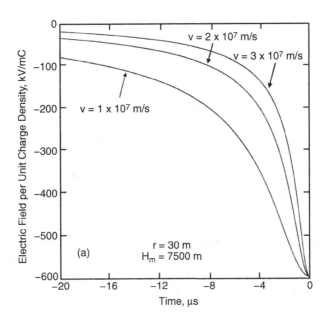

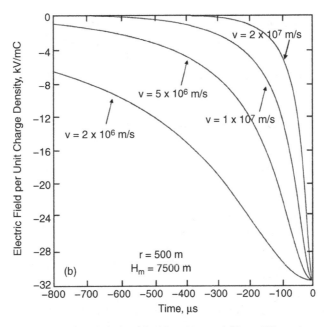

Fig. 4.15 Negative leader electric field waveforms calculated for (a) $r = 30$ m and (b) $r = 500$ m using a uniformly-charged vertical-line leader model with constant speed. The height of the charge source is assumed to be $H_m = 7.5$ km, although the close fields are relatively insensitive to the value of H_m. At $t = 0$ the leader attaches to the ground. In order to obtain kV m^{-1}, multiply the ordinate by $|\rho_L|$, the magnitude of line charge density in mC m^{-1}. The atmospheric electricity sign convention, according to which a downward-directed electric field or field change is defined as positive, is used here. Adapted from Rubinstein et al. (1995).

the observation point is very close to the channel base so that $r \ll H_m$, Eq. 4.7 can be further simplified to give

$$B_\phi(r, \ t) = \frac{\mu_0 I(t)}{2\pi r} \qquad (4.8)$$

which is the same equation as that for an infinitely long current-carrying wire in free space.

4.3.3 Leader to return stroke electric field change ratio

The ratio of the net leader to return stroke electric field changes is usually computed (1) assuming that the return stroke removes all the leader charge and does not deposit any additional charge either on the channel or in the cloud and (2) using a simple, uniformly charged leader model, which is illustrated in Fig. 4.13. With these assumptions, the leader electric field change is given by Eq. 4.4, and the return-stroke electric field change is given by the negative of the first two terms of Eq. 4.4. These two terms represent the contribution to the electric field change from the channel, while the third term accounts for the removal of that charge from the source (or, equivalently, the deposition of equal magnitude and opposite polarity charge at the source). The field ratio computed in this manner is shown as a function of H_m/r in Fig. 4.16. The curve crosses zero level at $H_m/r = \tan \alpha_s = 1.27$, where $\alpha_s = 52°$ is the angle between the line joining the source and the observation point and the ground surface, when the net leader field change is equal to zero. At relatively large ranges ($H_m/r < 1.27$), the leader and return stroke field changes have the same polarity, so that the ratio is positive and approaches unity at distances $r \gg H_m$. The latter result can be visualized as being due to the following two-stage process. In the first (leader) stage, an equivalent point charge equal to the total leader charge moves from its original position in

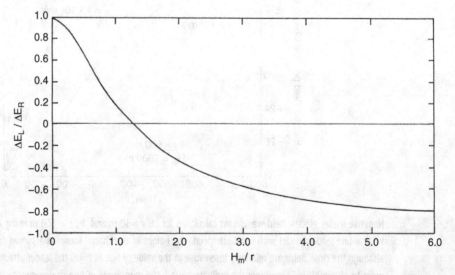

Fig. 4.16   The model-predicted ratio of the leader and return-stroke electric field changes as a function of the ratio of the height H_m of the charge center and the horizontal distance r from the charge center (r decreases from left to right). Adapted from Uman (1987, 2001).

the cloud to a lower position which, for the assumed uniform leader charge distribution, is exactly halfway between this original position and ground. In the second (return-stroke) stage, the equivalent charge moves from the intermediate position to the ground. At far ranges ($r \gg H_m$), these two stages of point charge lowering produce equal net electric field changes. At relatively close ranges ($H_m/r > 1.27$), the leader and return-stroke field changes have opposite polarities. The ratio is negative and approaches -1 at distances $r \ll H_m$. The latter result is due to the fact that the third, cloud-source term in Eq. 4.4 makes a negligible contribution to the leader electric field change at close ranges compared to the first two terms representing channel sources.

Experimental data on the ratio of the net leader to return stroke electric field changes along with model-based calculations are shown in Figs. 4.17 and 4.18. For first strokes (Figs. 4.17 and 4.18a) and new ground-termination subsequent strokes (Fig. 4.18b), there is

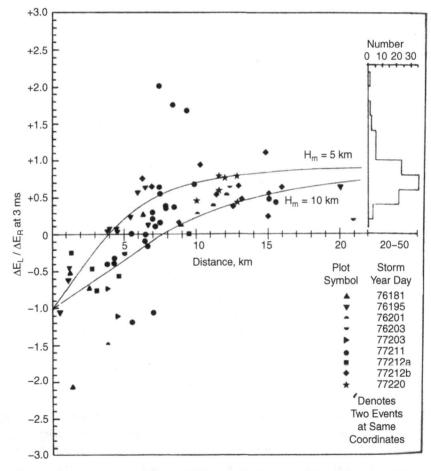

Fig. 4.17 Ratio of stepped-leader to return-stroke electric field changes for 80 leaders in nine Florida storms. The histogram in the upper right-hand side shows additional data for the distance range from 20 to 50 km. The upper and lower curves represent the model-predicted ratio for charge source heights $H_m = 5$ km and $H_m = 10$ km, respectively. Adapted from Beasley et al. (1982).

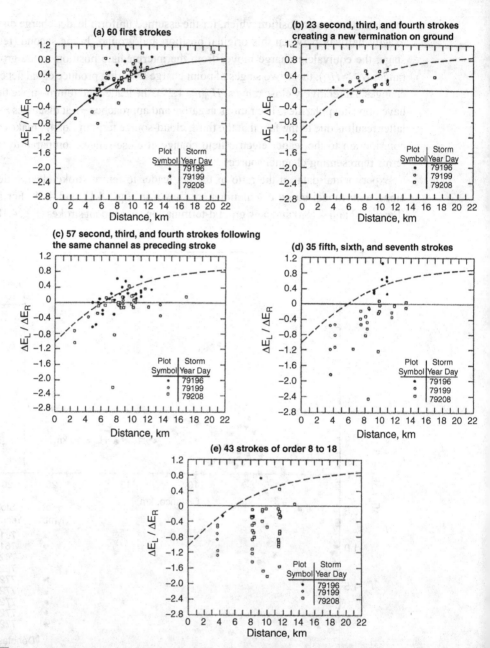

Fig. 4.18 Ratio of the leader and return-stroke electric field changes $\Delta E_L/\Delta E_R$ as a function of distance for five different sets of strokes: (a) 60 first strokes; (b) 23 second, third, and fourth strokes creating a new termination on ground; (c) 57 second, third, and fourth strokes following the same channel as the preceding stroke; (d) 35 fifth, sixth, and seventh strokes, and (e) 43 strokes of order 8 through 18. The data were obtained during three thunderstorms in Florida. Both the solid and broken curves (in (a) and (b)–(e), respectively) show the ratio predicted by a simple model having a vertical, uniformly charged channel and a cloud charge source centered at 7.5 km. Adapted from Rakov et al. (1990).

a fairly good agreement between model predictions and the experimental data, while for regular subsequent strokes (Figs. 4.18c–e) there is a significant mismatch which increases with stroke order. As another way of viewing the experimental data shown in Fig. 4.18, Rakov and Uman (1990c) reported that at distances ≥10 km 89 percent of first-stroke leaders exhibited monotonic positive electric field changes and none exhibited hook-shaped net negative field changes (as predicted by the model), while for subsequent strokes of order of 8 through 18 the proportion was exactly opposite. The disparity can be explained by assuming an inverted-L channel geometry for subsequent strokes, with the horizontal (in-cloud) channel section being predominantly oriented away from the observer and its length increasing with stroke order (Rakov et al., 1990).

The fairly good agreement for first strokes might be surprising in view of the simplicity of the model assuming a uniformly charged leader channel extending from a spherically symmetrical charge source. If the leader is represented by a vertical conductor extending in the cloud electric field, it will be polarized and its line charge density cannot be uniform. Nevertheless, for the histogram shown in Fig. 4.17 for relatively large distances, the mean leader to return stroke electric field change ratio is about 0.8, which is close to the simple model prediction for $r = 20$ km ($H_\mathrm{m} = 7.5$ km). Mazur and Ruhnke (1993) simulated the upper section of the bidirectional leader system by a single-channel positive leader extending vertically upward at the same speed as the downward-extending negative section of the leader system. They found that the charge density on such a vertically symmetrical bidirectional leader, assuming it was a vertical conductor polarized in a uniform electric field, varied linearly with height with zero charge density at the origin. The model of Mazur and Ruhnke (1993) predicts a ratio of leader-to-return-stroke electric field change at 20 to 50 km to be between approximately 0.2 and 0.3, which is significantly lower than the average value of 0.8 observed for first strokes at these distances by Beasley et al. (1982), while the leader model represented in Fig. 4.13 is consistent with those observations. It is worth noting that very close (within tens to hundreds of meters) electric field measurements for triggered lightning show that the uniform leader charge density distribution for the bottom kilometer or so of the channel is a reasonable approximation for computing fields on the ground (Crawford et al., 2001; Cooray et al., 2004).

4.3.4 Leader steps

The step-formation process in lightning occurs on a timescale of less than 1 μs or so. As a result, it is difficult to adequately resolve this process in optical records of lightning. On the other hand, there appears to be a qualitative similarity between the lightning negative stepped leader and the negative stepped leader of the long laboratory spark. The latter type of leader is much better studied via the use of electronic image-converter cameras in conjunction with the measurement of current through the air gap. The negative long spark leader exhibits distinct steps when the gap length is several meters or more. Gorin et al. (1976), for example, reported that a 6 m rod-plane gap was bridged by a negative leader in three to five steps. It is worth noting that stepping is observed in lightning negative leaders regardless of whether they are initiated in the cloud (downward leaders) or at the grounded object (upward leaders). This fact suggests that the mechanism of formation of a

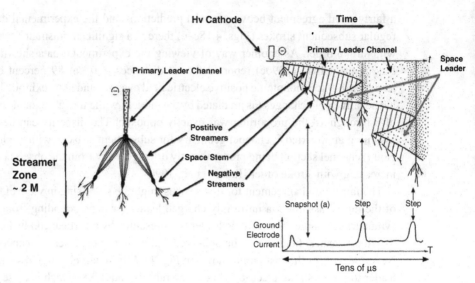

Fig. 4.19   Illustration of the development of negative stepped leader in a long laboratory spark, based on a description given by Gorin et al. (1976). It schematically shows a snapshot (left) and a time-resolved optical picture (upper right) including an initial impulsive corona from the negative high-voltage electrode and the first two steps, along with the corresponding current through the gap (lower right). Adapted from Biagi et al. (2010).

step is determined primarily by the processes at the leader tip and in the leader channel rather than being determined by the source (cloud charge for lightning and impulse generator circuitry for long laboratory sparks).

The development of the negative stepped leader in a long laboratory spark, based on a description given by Gorin et al. (1976), is illustrated in Fig. 4.19 which schematically shows a snapshot (left) and a time-resolved optical picture (upper right) including an initial impulsive corona from the negative high-voltage electrode and the first two steps, along with the corresponding current through the gap (lower right). The snapshot shows the primary leader channel and three streamer branches that are in contact with the leader tip. Each streamer branch has a plasma formation, termed space stem, and positive (upward-directed) and negative (downward-directed) streamers. The initial impulsive corona, a system of branched filamentary channels seen in the time-resolved picture, serves to heat the air near the high-voltage electrode and to form the initial section of the leader plasma channel. This process produces the first pulse in the current record (lower right), the other two current pulses being associated with the two leader steps. The initial section of the leader channel extends from the high-voltage electrode into the gap. The leader tip is brighter than the channel behind it and is shown as a slightly curved, negatively sloped solid line. The positive streamers develop toward the leader tip (these are shown in the time-resolved picture by longer, positively sloped solid lines), and the negative streamers develop into the gap (these are shown by shorter, negatively sloped lines). The oppositely charged streamers start from the space stem (shown as a negatively sloped dashed line), which moves into the gap in front of the leader tip. The air behind the moving space stem apparently remains essentially an insulator. It is presently unknown how a space stem is formed ahead of the leader tip. When the space stem is sufficiently heated, it gives rise to, in effect, a section of the leader channel that is not connected to either of the electrodes and extends in both upward and downward directions. The bidirectional channel extension is shown in Fig. 4.19 (upper right) as a pair of slightly curved and diverging solid lines drawn from a common origin on the slanted dashed line. The upward-extending part of this bidirectional space leader is charged positively, and the downward-extending part is charged negatively. The first leader step is formed at the instant when the upward-moving positive end of the space-leader channel makes contact with the downward-moving negative tip of the primary (that is, connected to the high-voltage electrode) leader channel. At that moment, the very high potential (close to the potential of the high-voltage electrode) of the primary leader channel is rapidly transferred to the lower end of the space-leader channel. As a result, a burst of negative streamers (thought to be responsible for production of X-rays by negative leaders) is produced at the bottom of the newly added channel section. Such a breakdown generates a current pulse which propagates toward the high-voltage electrode and briefly illuminates the entire channel. Thus, in the negative leader, the occurrence of each luminous step is caused by the connection of a space leader to the primary leader channel. The development of the next step of a negative leader begins with the formation of a new space stem ahead of the newly added leader step.

Biagi et al. (2010), using a high-speed video camera (4.17 μs frame integration time), imaged the bottom 150 m of a downward negative, dart-stepped leader in a rocket-and-wire triggered flash. They observed vertically elongated luminous formations, 1–4 m in length, which were separated by darker gaps, 1–10 m in length, from the bottom of the downward-extending leader channel, shown in Fig. 4.20. These formations, indicated by white arrows in seven of the nine frames, are similar to the space stems or space leaders that have been imaged in long negative laboratory sparks. It is worth noting that the frame integration time (about 4 μs) was longer than the duration of step-formation process (less than 1 μs), so that individual steps could not be resolved.

Wang et al. (1999b), using a high-speed (100 ns sampling interval) digital optical imaging system (ALPS) in rocket-triggered lightning experiments at Camp Blanding, Florida, reported observations of luminosity waves that originated at newly formed leader steps in a dart-stepped leader and propagated toward the cloud, as illustrated in Fig. 4.21. Shown in this figure are leader light versus time waveforms at eight different heights above ground. The overall downward progression of the leader is shown by large, negatively sloped arrow. The light pulses corresponding to individual steps are numbered. These pulses appear to move upward, as indicated by dotted, positively sloped line for pulse No. 10, and attenuate to about 10 percent of the original luminosity value after traveling 50 m or so. The dotted line is considerably steeper than the large negatively sloped arrow, which means that the upward light pulse speed (mean = 6.7×10^7 m/s) is much higher than the downward leader progression speed (2.5×10^6 m/s, on average). Note that the spatial resolution of ALPS was about 30 m and, because of that, not all the individual step pulses were resolved (more than one step could be formed within 30 m).

In optical records, each lightning leader step is typically tens of meters in length, with a time interval between steps of 20 to 50 μs. The peak value of the current pulse associated

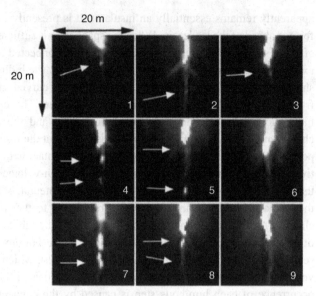

Fig. 4.20   The bottom 20 m of the downward-extending leader channel of a triggered-lightning flash (Camp Blanding, Florida) in the nine high-speed video frames (240 kfps, 4.17 μs per frame). Each image shows about 20 m × 20 m. The white arrows point to the luminous segments (space stems or leaders), 1–4 m in length, that formed separately from and 1–10 m below the downward-extending leader channel. The leader traveled about 100 m from frame 1 to frame 9 where it was about 30 m above its termination point. The return stroke began during frame 10. Adapted from Biagi et al. (2010).

with an individual step is at least 1 kA, and the minimum charge is a few millicoulombs. Examples of electric field pulses associated with negative lightning leader steps are shown in Fig. 4.22.

4.4 Attachment process

The process of lightning leader attachment to ground or to a grounded object is one of the least understood and poorly documented processes of the cloud-to-ground lightning discharge. It is often assumed that the attachment process begins when an upward-moving leader is initiated in response to the approaching downward-moving leader at the ground or, more likely, at the tip of an object protruding above ground. It is possible that two or more upward leaders are launched from the ground toward the descending leader, perhaps in response to different branches of the descending leader. An upward leader that makes contact with a branch of a downward leader is called an "upward connecting leader." The so-called breakthrough phase is assumed to begin when the relatively low conductivity streamer zones ahead of the two propagating leader tips meet to form a common streamer zone. The subsequent accelerated extension of the two relatively high conductivity plasma

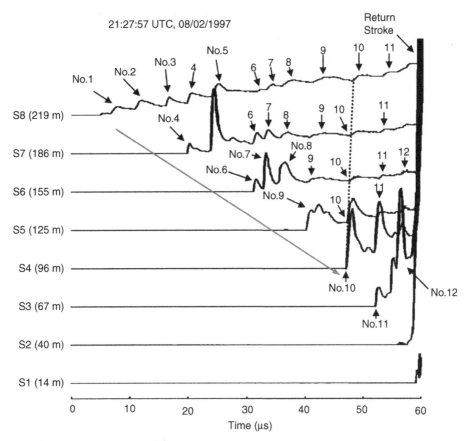

Fig. 4.21 Light versus time waveforms at different heights above ground for a dart-stepped leader in a negative flash triggered using the rocket-and-wire technique at Camp Blanding, Florida. Large, negatively sloped arrow indicates the overall downward progression of the leader. Light pulses associated with individual steps are numbered. They appear to originate at the tip of the downward-extending leader channel and propagate upward. Adapted from Wang et al. (1999b).

channels toward each other takes place inside the common streamer zone. The breakthrough phase can be viewed as a switch-closing operation that serves to launch two return-stroke waves from the point of junction between the two plasma channels. One wave moves downward, toward the ground, and the other upward, toward the cloud. The downward-moving return-stroke wave quickly reaches the ground, and the resultant upward reflected wave from ground catches up with the upward-moving return-stroke wave. The latter is the case because the reflected wave from the ground propagates in the return-stroke-conditioned channel and, hence, is likely to move faster than the upward wave from the junction point that propagates along the leader-conditioned channel. Eventually, a single upward-moving wave is formed. Thus, the lightning attachment process involves two plasma channels growing toward each other, initially in air (the upward connecting leader phase) and then inside the

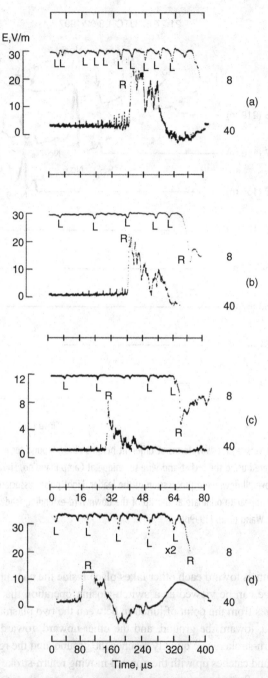

Fig. 4.22 Electric field waveforms produced by four negative first strokes in Florida at distances of some tens of kilometers. Each record is displayed on two timescales, 8 μs per division (upper trace) and 40 μs per division (lower trace), the two traces being inverted with respect to each other. Note that the abrupt return-stroke transitions (clipped in the lower trace of Fig. 4.22d), labeled R, are preceded by small pulses characteristic of leader steps, each labeled L. The vertical scale is shown on the left (this is to be reduced by a factor of 2 for the 8 μs per division (upper) trace in Fig. 4.22d). Adapted from Krider et al. (1977).

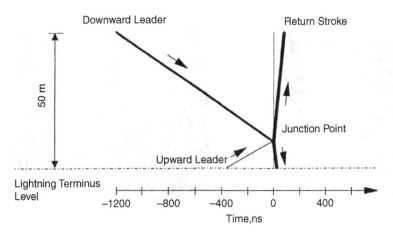

Fig. 4.23 Schematic representation of upward connecting leader and bidirectional return-stroke process observed in rocket-triggered lightning at Camp Blanding, Florida. Adapted from Wang et al. (1999a).

common streamer zone (the breakthrough phase). It is a matter of definition whether the very short-lived bidirectional return-stroke wave should be considered a part of the attachment process or a part of the return stroke.

Wang et al. (1999a), using the digital optical imaging system ALPS with 3.6 m spatial and 100 ns time resolution, observed an upward connecting leader (but not the break-through phase) in one triggered-lightning stroke and inferred the existence of such a leader in another one. A sketch of the time-resolved image for the former event is shown in Fig. 4.23, in which the tips of the descending and upward connecting leaders come in contact at $t = 0$. In both events, the return stroke was initially a bidirectional process that involved both upward- and downward-moving waves which originated at 7–11 m (in the event with the imaged upward connecting leader) and 4–7 m (in the event with no imaged upward connecting leader). The propagation speed of the upward con-necting leader was estimated to be about 2×10^7 m/s, similar to the typical speed of downward dart leaders.

Biagi et al. (2009), using a high-speed video camera (20 μs frame integration time), recorded upward connecting leaders, ranging from 9 to 22 m in length, in eight strokes of a single rocket-triggered lightning flash. One of these upward leaders, whose length was 16 m, is shown in Fig. 4.24a. This figure also shows a faint streamer filament between the downward and upward connecting leaders, apparently indicative of the beginning of the breakthrough phase. Similar features can be seen in optical images of two long negative laboratory sparks at different stages of their development, shown in Figs. 4.24b and c. In Fig. 4.24b, L_s is the length of just-formed step with a pronounced burst of negative streamers from its lower end. The streamer burst signifies the final stage of step-formation process (see Section 4.3). In Fig. 4.24b, there is a faint streamer filament bridging the gap between the downward and upward leaders, similar

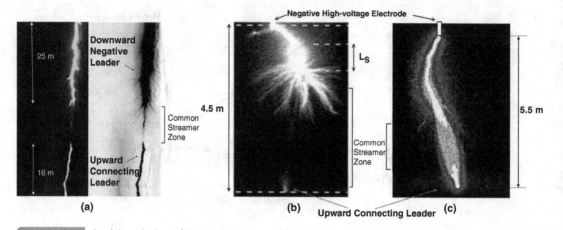

Fig. 4.24   Breakthrough phase of the attachment process in rocket-triggered lightning and in long laboratory sparks. (a) shows the original and inverted-intensity versions of the high-speed video frame (20 μs integration time) just prior to the return stroke frame of a negative triggered-lightning stroke. (b) and (c) are single frames of two negative laboratory sparks captured by image converter camera K008 with frame durations of 2 μs and 0.2 μs, respectively. L_s in (b) is the length of just-formed step (note the burst of negative streamers from the lower end of the step) of the downward negative leader. In (c), light intensity is color coded, with the highest intensity shown in white. The common streamer zone (a faint filament in (b), which is difficult to see in a reproduction) between the downward leader channel and the upward connecting leader channel is imaged in all three cases. Images shown in (a), (b), and (c) are adapted from Biagi et al. (2009), Lebedev et al. (2007), and Shcherbakov et al. (2007), respectively. The figure was compiled and annotated by Manh D. Tran.

to that seen in Fig. 4.24a. In Fig. 4.24c, light intensity is color coded, with the highest intensity shown in white.

4.5 Return stroke

The return stroke has been the most studied lightning process due to both practical considerations (it is the return-stroke current that is thought to produce most of the damage attributable to lightning) and the fact that of all the processes comprising a lightning flash, the return stroke lends itself most easily to measurement. Indeed, the return stroke is the optically brightest lightning process visible outside of the cloud, and it produces the most readily identifiable wideband electromagnetic signature.

In this section, we present characterization of negative return strokes, including (1) parameters derived from channel-base current measurements (Section 4.5.1), (2) luminosity variation along the channel (Section 4.5.2), (3) propagation speed (Section 4.5.3), equivalent impedance of the lightning channel (Section 4.5.4), (4) electric and magnetic fields (Section 4.5.5), and peak currents inferred from measured fields (Section 4.5.6). The calculation of electric and magnetic fields is considered in Chapter 5. Models of the return stroke that are used in field calculations are discussed in Chapter 6.

4.5.1 Parameters derived from channel base current measurements

The most complete characterization of the return stroke in negative downward flashes is due to Karl Berger and co-workers (e.g. Berger, 1972; Berger et al., 1975). The data of Berger were derived from oscillograms of current measured using resistive shunts installed at the tops of two 70 m high towers on the summit of Monte San Salvatore in Lugano, Switzerland. The summit of the mountain is 915 m above sea level and 640 m above the level of Lake Lugano, located at the base of the mountain. The towers are of moderate height, but because the mountain contributed to the electric field enhancement near the tower tops, the effective height of each tower was a few hundred meters. As a result, the majority of lightning strikes to the towers were of the upward type. Here we only consider return strokes in negative downward flashes. A total of 101 are included in the summary by Berger et al. (1975). Berger's data were additionally analyzed by Anderson and Eriksson (1980).

The results of Berger et al. (1975) are still used to a large extent as the primary reference source for both lightning protection and lightning research. These results are presented in Figs. 4.25 and 4.26 and in Table 4.4. Figure 4.25 shows, on two timescales, A and B, the average current waveshapes for negative first and subsequent strokes. The averaging procedure involved the normalization of waveforms from many strokes to their respective peak currents (so that all have peaks equal to unity) and subsequent alignment using the 0.5 peak point on the initial rising portion of the waveforms. The overall duration of the current waveforms is some hundreds of microseconds. The rising portion of the first-stroke waveform has a characteristic concave shape. The averaging procedure masked secondary maxima typically observed in first-stroke waveforms and generally attributed to major branches. Figure 4.26 shows the cumulative statistical distributions (solid-line curves) of return-stroke peak currents for (1) negative first strokes, (2) negative subsequent strokes, and (3) positive strokes (each was the only stroke in a flash). These empirical distributions are approximated by log-normal distributions (dashed lines) and shown on cumulative probability distribution graph paper, on which a Gaussian (normal) cumulative distribution appears as a slanted straight line, with the horizontal (peak current) scale being logarithmic (base 10). The vertical scale gives the percentage of peak currents exceeding a given value on the horizontal axis. The vertical scale is symmetrical with respect to the 50-percent value and does not include the 0- and 100-percent values; it only asymptotically approaches those. For a log-normal distribution the 50-percent (median) value is equal to the geometric mean value. The lightning peak current distributions for negative first and subsequent strokes shown in Fig. 4.26 are also characterized by their 95-, 50-, and 5-percent values based on the log-normal approximations in Table 4.4, which contains a number of other parameters derived from the current oscillograms. The minimum peak current value included in the

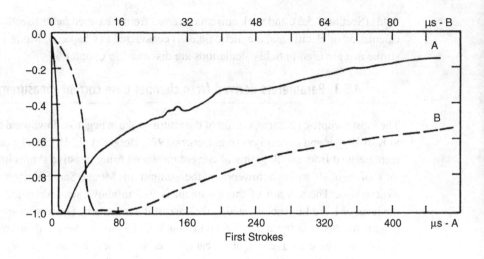

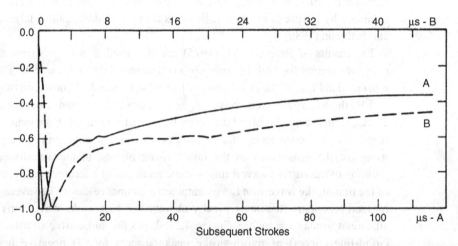

Fig. 4.25 Average negative first- and subsequent-stroke current waveshapes each shown on two timescales, A and B. The lower timescales (A) correspond to solid curves, while the upper timescales (B) correspond to broken curves. The vertical (amplitude) scale is in relative units, the peak values being equal to negative unity. Adapted from Berger et al. (1975).

distributions is 2 kA, although no first strokes (of either polarity) with peak currents below 5 kA were observed.

Clearly, the parameters of statistical distributions can be affected by the lower and upper measurement limits. For a log-normal distribution, the parameters of a measured, "truncated" distribution and a knowledge of the lower measurement limit can be used to recover the parameters of the actual, "untruncated" distribution. Rakov (1985) applied the recovery procedure to the various lightning peak current distributions found in the literature and

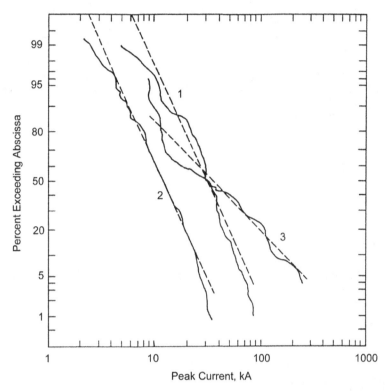

Fig. 4.26 Cumulative statistical distributions of return-stroke peak current (solid curves) and their log-normal approximations (broken lines) for (1) negative first strokes, (2) negative subsequent strokes, and (3) positive first (and only) strokes, as reported by Berger et al. (1975).

concluded that the peak current distributions published by Berger et al. (1975) can be viewed as practically unaffected by the effective lower measurement limit of 2 kA. Further, it has been shown by Rakov (2003) that Berger's peak currents for negative first strokes, based on measurements at the top of 70 m towers, are not much influenced by the transient process (reflections) excited in the tower. For negative subsequent strokes, reflections are expected to increase the tower-top current by only 10 percent or so. The distribution of peak currents based on measurements on tall instrumented towers may be biased (relative to the ground-surface peak-current distribution) toward higher values due to the peak-current-dependent attractive effect of the tower, but, as of today, there is no experimental evidence of this effect. To summarize, it appears that Berger's distributions of peak currents for first and subsequent negative strokes can be viewed as not materially affected by either lower measurement limit or the presence of the tower.

Berger's peak current distributions for first and subsequent negative strokes are generally confirmed by more recent direct current measurements, particularly those with larger sample sizes obtained in Japan (first strokes, $N = 120$; Takami and Okabe 2007), Austria (subsequent strokes, $N = 615$; Diendorfer et al. 2009), and Florida (subsequent strokes, $N = 165$; Schoene et al. 2009). At the same time, direct current measurements in Brazil

Table 4.4 Parameters of downward negative lightning derived from channel-base current measurements. Adapted from Berger et al. (1975)

Parameters	Units	Sample Size	Percent Exceeding Tabulated Value		
			95%	50%	5%
Peak current (minimum 2 kA)	kA				
First strokes		101	14	30	80
Subsequent strokes		135	4.6	12	30
Charge (total charge)	C				
First strokes		93	1.1	5.2	24
Subsequent strokes		122	0.2	1.4	11
Complete flash		94	1.3	7.5	40
Impulse charge (excluding continuing current)	C				
First strokes		90	1.1	4.5	20
Subsequent strokes		117	0.22	0.95	4
Front duration (2 kA to peak)	µs				
First strokes		89	1.8	5.5	18
Subsequent strokes		118	0.22	1.1	4.5
Maximum dI/dt	kA µs^{-1}				
First strokes		92	5.5	12	32
Subsequent strokes		122	12	40	120
Stroke duration (2 kA to half-peak value on the tail)	µs				
First strokes		90	30	75	200
Subsequent strokes		115	6.5	32	140
Action integral ($\int I^2 dt$)	A^{2}s				
First strokes		91	6.0×10^3	5.5×10^4	5.5×10^5
Subsequent strokes		88	5.5×10^2	6.0×10^3	5.2×10^4
Time interval between strokes	ms	133	7	33	150
Flash duration	ms				
All flashes		94	0.15	13	1100
Excluding single-stroke flashes		39	31	180	900

(Visacro et al. 2012) yielded 50 percent higher median peak currents for both first ($N = 38$) and subsequent ($N = 71$) strokes.

In lightning protection standards, in order to increase the sample size, Berger's data are often supplemented by limited direct current measurements in South Africa and by less accurate indirect lightning current measurements obtained in different countries using magnetic links (ferromagnetic recorders of peak current). There are two main distributions of lightning peak currents for negative first strokes adopted by lightning protection standards: the IEEE distribution (e.g. IEEE Std 1410–2010; IEEE Std 1243–1997) and CIGRE distribution (e.g. Anderson and Eriksson, 1980; CIGRE Document 63, 1991). [CIGRE is a French acronym which stands for the International Council on Large Electric Systems, and

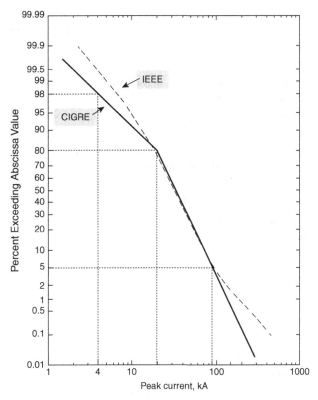

Fig. 4.27   Cumulative statistical distributions of peak currents for negative first strokes adopted by IEEE and CIGRE and used in various lightning protection standards. Adapted from CIGRE Document 63 (1991). Drawing by Yongfu Li.

IEEE stands for the Institute of Electrical and Electronics Engineers.] Both these "global distributions" are presented in Fig. 4.27). In the coordinates of Fig. 4.27 (same as in Fig. 4.26), a cumulative log-normal distribution appears as a slanted straight line. The CIGRE peak current distribution is approximated by two slanted lines having different slopes and intersecting at 20 kA. For this distribution, 98 percent of peak currents exceed 4 kA, 80 percent exceed 20 kA, and 5 percent exceed 90 kA. For the IEEE distribution, the "probability to exceed" values are given by the following equation:

$$P(I) = \frac{1}{1 + (I/31)^{2.6}} \qquad (4.9)$$

The peak-current distribution for subsequent strokes adopted by the IEEE (IEEE Std 1243–1997; IEEE Std 1410–2010) is given by

$$P(I) = \frac{1}{1 + (I/12)^{2.7}} \qquad (4.10)$$

CIGRE recommends for negative subsequent strokes a log-normal distribution of peak currents with the median of 12.3 kA and the logarithmic (base e) standard deviation of 0.53 (CIGRE Document 63 (1991)).

It follows from Fig. 4.26 and Table 4.4 that the median return-stroke current peak for first strokes is two to three times higher than that for subsequent strokes. Also, negative first strokes transfer a charge of about a factor of 4 greater than those of negative subsequent strokes. On the other hand, subsequent return strokes are characterized by a three to four times higher current maximum steepness (current maximum rate of rise). It is important to note that the current maximum rate of rise reported by Berger et al. (1975) and given in Table 4.4 is an underestimate of the actual value due to the limited time resolution of oscillographic data, as further discussed below. Only a few percent of negative first strokes are expected to exceed 100 kA, while about 20 percent of positive strokes have been observed to do so. On the other hand, the 50-percent (median) values of the current distributions for negative first and positive strokes are similar. The action integral (also referred to as specific energy) in Table 4.4 represents the energy that would be dissipated in a 1 Ω resistor if the lightning current were to flow through it. It is thought that the heating of electrically conducting materials and the explosion of nonconducting materials is, to a first approximation, determined by the value of the action integral. Note that the inter-stroke interval in Table 4.4 is likely mislabeled by Berger et al. (1975) and is actually the no-current interval: that is, the inter-stroke interval excluding any continuing current.

The median 10-to-90 percent rise time estimated for subsequent strokes by Anderson and Eriksson (1980) from Berger et al.'s (1975) oscillograms is 0.6 μs, comparable to the median values ranging from 0.3 to 0.6 μs for triggered lightning strokes (Leteinturier et al., 1991; Fisher et al., 1993). The median 10-to-90 percent current rate of rise reported for natural subsequent strokes by Anderson and Eriksson (1980) is 15 kA/μs, almost three times lower than the corresponding value of 44 kA/μs in data of Leteinturier et al. (1991) and more than twice lower than the value of 34 kA/μs found by Fisher et al. (1993). The mean value of current derivative peak reported by Leteinturier et al. (1991) is 110 kA/μs. The higher observed values of current rate of rise for triggered-lightning return strokes than for natural return strokes are likely due to the use of modern instrumentation (digital oscilloscopes with better upper frequency response). Triggered-lightning data for current rates of rise can be applied to subsequent strokes in natural lightning.

For triggered lightning, relatively strong correlation is observed between the lightning peak current and impulse charge transfer and between the current rate-of-rise character-istics and current peak, and relatively weak or no correlation between the peak and rise time and between the peak and half-peak width.

4.5.2 Luminosity variation along the channel

The variation of return-stroke luminosity along the channel is thought to reflect the variation of current. Since the latter variation is impossible to measure directly, the luminosity profile along the channel, typically obtained using streak photography, is generally viewed as representative of current variation. The light profile along a subsequent return stroke channel usually shows a gradual intensity decay with height as illustrated in

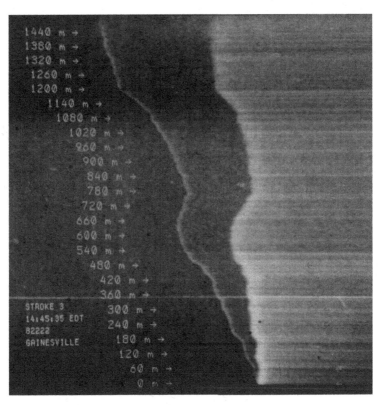

Fig. 4.28a Streak photograph of stroke 3 in a flash that occurred in Gainesville, Florida. The image on the left is due
to the dart leader, and the brighter image on the right is due to the return stroke. Adapted from Jordan et al. (1997).

Figs. 4.28a–c. As seen in Figs. 4.28b and c, besides a decrease in peak value, the return-stroke light pulses also exhibit an appreciable increase in rise time (that is, degradation of the front) with increasing height.

4.5.3 Propagation speed

Lightning return stroke speed is a parameter in the models used in evaluating lightning-induced effects in power and communication lines. Further, an explicit or implicit assumption of the return-stroke speed is involved in inferring lightning currents from remotely measured electric and magnetic fields (see Section 4.5.6). It is known that the return-stroke speed may vary significantly along the lightning channel. As a result, optical speed measurements along the entire channel are not necessarily representative of the speed within the bottom 100 m or so, that is, at early times when the peaks of the channel-base current and of remote electric and magnetic fields are formed.

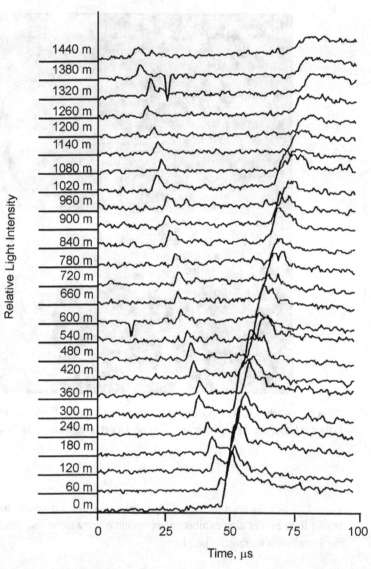

Fig. 4.28b   Luminosity versus time at different heights above ground corresponding to the streak photograph in Fig. 4.28a.

The optically measured return-stroke speed probably represents the speed of the region of the upward-moving return-stroke tip where the input power is greatest, the power per unit length being the product of the current and the longitudinal electric field in the channel. The peak of the input power (Joule heating) wave likely occurs earlier in time than the peak of the current wave. Since the shape of the return-stroke light pulse changes significantly with height, there is always some uncertainty in tracking the propagation of such pulses for a speed measurement. For example, if the light pulse peak is tracked, then an increase in pulse

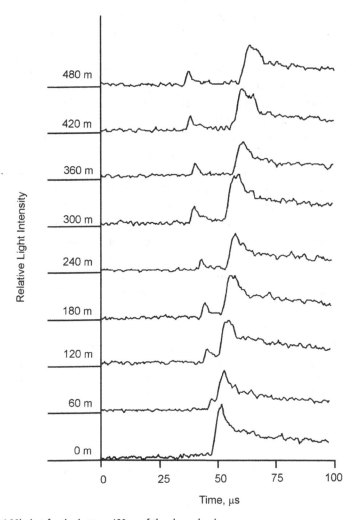

Fig. 4.28c Same as Fig. 4.28b, but for the bottom 480 m of the channel only.

rise time translates into a lower speed value than if an earlier part of the light pulse is tracked. Techniques for measuring return-stroke speed are discussed, for example, by Idone and Orville (1982) and Olsen et al. (2004).

The average propagation speed of a negative return stroke (first or subsequent) below the lower cloud boundary is typically between one-third and one-half of the speed of light. From streak-camera measurements of Idone and Orville (1982), the return-stroke speed for first strokes is somewhat lower than that for subsequent strokes, although the difference is not very large (9.6×10^7 vs. 1.2×10^8 m/s). The speed within the bottom 100 m or so is expected to be between one-third and two-thirds of the speed of light. The negative return stroke speed usually decreases with height for both first and subsequent strokes. There exists some experimental evidence that the negative return stroke speed may vary non-monotonically along the lightning channel, initially increasing and then decreasing with increasing height. More detailed information about the return-stroke speed can be found in Rakov (2007) and in references therein.

We now discuss why the return-stroke speed is lower than the speed of light. It is well known that the propagation speed of waves on a uniform, linear, and lossless transmission line surrounded by air is equal to the speed of light, $c = (LC)^{-1/2} = (\mu_0\varepsilon_0)^{-1/2}$. However, a vertical lightning channel (leaving aside the general validity of its transmission-line approximation) is a nonuniform, nonlinear, and lossy transmission line. Indeed, its inductance and capacitance per unit length vary with height above ground, so that its characteristic impedance, $(L/C)^{1/2}$, increases with height. Thus, a return-stroke wave will suffer distortion even in the absence of losses. Further, charge cannot be confined within the narrow channel core carrying the longitudinal current; it is pushed outward via electrical breakdown forming the radial corona sheath. Finally, channel resistance per unit length ahead of the return-stroke front is relatively high (causing wave attenuation and dispersion) and decreases by two orders of magnitude or so behind the front.

Two primary reasons for the lightning return-stroke speed, v, being lower than the speed of light are: (1) the effect of radial corona surrounding the narrow channel core (the radius of the charge-containing corona sheath is considerably larger than the radius of the core carrying the longitudinal channel current, so that $(LC)^{-1/2} < \mu_0\varepsilon_0^{-1/2} = c$, and hence $v < c$), and (2) the ohmic losses in the channel core that are usually represented in lightning models by the distributed constant or current-dependent series resistance of the channel.

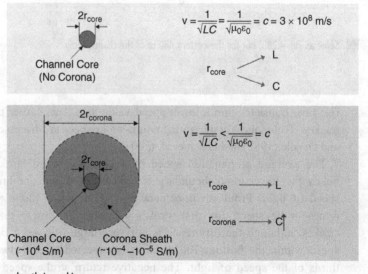

Role of corona in making the return-stroke speed lower than the speed of light

$$v = \frac{1}{\sqrt{LC}} = \frac{1}{\sqrt{\mu_0\varepsilon_0}} = c = 3 \times 10^8 \text{ m/s}$$

2r_core

Channel Core
(No Corona)

r_{core} → L
r_{core} → C

2r_corona

2r_core

$$v = \frac{1}{\sqrt{LC}} < \frac{1}{\sqrt{\mu_0\varepsilon_0}} = c$$

r_{core} ⟶ L
r_{corona} ⟶ C↑

Channel Core   Corona Sheath
(~10^4 S/m)   (~10^{-4}–10^{-5} S/m)

Fig. 4.29   Role of corona sheath in making $v < c$.

The corona effect explanation is illustrated in Fig. 4.29. It is based on the following assumptions:

(a) The longitudinal channel current flows only in the channel core, because the core conductivity, of the order of 10^4 S/m, is much higher than the corona sheath conductivity, of the order of 10^{-6}–10^{-5} S/m. The longitudinal resistance of channel core is expected to be about 3.5 Ω/m, while that of a 2 m radius corona sheath should be of the order of kiloohms to tens of kiloohms per meter. The corona current is radial (transverse) and hence cannot influence the inductance of the channel.

(b) The radial voltage drop across the corona sheath is negligible compared to the potential of the lightning channel. The average radial electric field within the corona sheath is about 0.5–1.0 MV/m, which results in a radial voltage drop of 1–2 MV across a 2 m radius corona sheath (expected for subsequent return strokes). The typical channel potential (relative to reference ground) is about 10–15 MV for subsequent strokes. For first strokes, both the corona sheath radius and channel potential are expected to be larger, so that about an order of magnitude difference between the corona sheath voltage drop and channel potential found for subsequent strokes should hold also for first strokes.

(c) The magnetic field due to the longitudinal current in channel core is not significantly influenced by the corona sheath. For corona sheath conductivity of 10^{-6}–10^{-5} S/m and frequency of 1 MHz, the field penetration depth is 160 to 500 m (and more for lower frequencies), which is much larger than the radius of corona sheath expected to be, of the order of meters.

In summary, the corona sheath conductivity is low enough to neglect both the longitudinal current through the sheath and the shielding effect of the sheath, but high enough to disregard the radial voltage drop across the sheath.

4.5.4 Equivalent impedance of the lightning channel

The equivalent impedance of the lightning channel is needed for specifying the source in circuit models used in studies of either direct-strike or induced lightning effects.

When the actual current distribution in the lightning channel is of no concern (electromagnetic coupling effects are neglected), lightning is often approximated by a Norton equivalent circuit. This representation includes an ideal current source equal to the lightning return stroke current that would be injected into the ground if that ground were perfectly conducting (a short-circuit current, I_{sc}) in parallel with a lightning-channel impedance Z_{ch} assumed to be constant. In the case when the strike object can be represented by lumped grounding impedance, Z_{gr}, this impedance is a load connected in parallel with the lightning Norton equivalent (see Fig. 4.30a). Thus, the "short-circuit" lightning current I_{sc} effectively splits between Z_{gr} and Z_{ch} so that the current flowing from the lightning-channel base into the ground is found as $I_{gr} = I_{sc}Z_{ch}/(Z_{ch} + Z_{gr})$. Both source characteristics, I_{sc} and Z_{ch}, vary from stroke to stroke, and Z_{ch} is a function of channel current, the latter nonlinearity being in violation of the linearity requirement necessary for obtaining the Norton equivalent

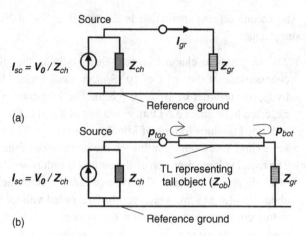

Fig. 4.30 Engineering models of lightning strikes (a) to lumped grounding impedance and (b) to a tall grounded object, in which lightning is represented by the Norton equivalent circuit, labeled "Source." Adapted from Baba and Rakov (2005a).

circuit. Nevertheless, Z_{ch}, which is usually referred to as equivalent impedance of the lightning channel, is assumed to be constant.

Equivalent impedance of the lightning channel also influences the transient process in the strike object, if that object is electrically long; that is, has dimensions that are comparable to or greater than the shortest significant wavelength of the source current, which is usually associated with the initial rising portion of the return-stroke current waveform. As an example, representation of lightning by a Norton equivalent circuit for analyzing lightning interaction with a tall object is shown in Fig. 4.30b.

It is possible to estimate the equivalent impedance of the lightning channel from measurements of lightning current waveforms at a very tall object, if the characteristic impedance of the object and the grounding impedance are known or can be reasonably assumed. Such measurements were performed at the 540 m high Ostankino tower in Moscow. Typical lightning current waveforms, reported by Gorin and Shkilev (1984), at heights of 47, 272, and 533 m above ground are shown in Fig. 4.31. The median peak currents from their measurements at 47 and 533 m were 18 and 9 kA, respectively. The observed differences in peak current suggest that the effective grounding impedance of the tower is much smaller than its characteristic impedance and that the latter impedance is appreciably lower than the equivalent impedance of lightning channel. Gorin and Shkilev (1984) used current oscillograms recorded near the tower top (at 533 m), for the cases in which the current rise time was smaller than the time (about 3.5 µs) required for a current wave to travel at the speed of light from the tower top to its base and back, to estimate the equivalent impedance of the lightning channel, assumed to be a real number. Their estimates varied from 600 Ω to 2.5 kΩ, when the characteristic impedance of the tower was assumed to be 300 Ω and the grounding resistance was assumed to be zero (the low-current, low-frequency value was about 0.2 Ω).

Thus, the estimates of the equivalent impedance of the lightning channel from limited experimental data suggest values ranging from several hundred ohm to a few kiloohm. In

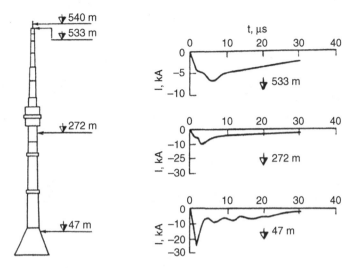

Typical return-stroke current waveforms of upward negative lightning recorded near the top (at 533 m), in the middle (at 272 m), and near the bottom (at 47 m) of the 540 m high Ostankino tower in Moscow. Differences in current waveforms at different heights are indicative of the tower behaving as a distributed circuit. Adapted from Gorin and Shkilev (1984).

many practical situations the impedance "seen" by lightning at the strike point is some tens of ohm or less, which allows one to assume infinitely large equivalent impedance of the lightning channel. In other words, lightning in these situations can be viewed as an ideal current source. In the case of direct lightning strike to an overhead conductor of a power line with 400 Ω surge impedance (effective impedance 200 Ω, since 400 Ω is "seen" in either direction), the ideal current source approximation may still be suitable.

4.5.5 Electric and magnetic fields

In this section, we will discuss measured waveforms of the vertical electric and horizontal magnetic fields produced at ground level by negative return strokes on microsecond and submicrosecond timescales. Return-stroke vertical electric fields on a millisecond time-scale are found in Figs. 4.4, 4.7, 4.8, 4.11, and 4.12. Typical vertical electric and horizontal (azimuthal) magnetic field waveforms at distances ranging from 1 to 200 km for both first and subsequent strokes have been published by Lin et al. (1979). These waveforms, which are drawings based on many measurements acquired in Florida, are reproduced in Fig. 4.32.

The electric fields of strokes observed within a few kilometers of the flash, shown in Fig. 4.32, are, after the first few tens of microseconds, dominated by the electrostatic component of the total electric field, the only field component which is nonzero after the stroke current has ceased to flow. The individual field components that comprise the electric and magnetic fields are discussed in Chapter 5. The close magnetic fields at similar times are dominated by the magnetostatic component of the total magnetic field, the component that produces the magnetic field humps seen in Fig. 4.32. Distant electric and magnetic

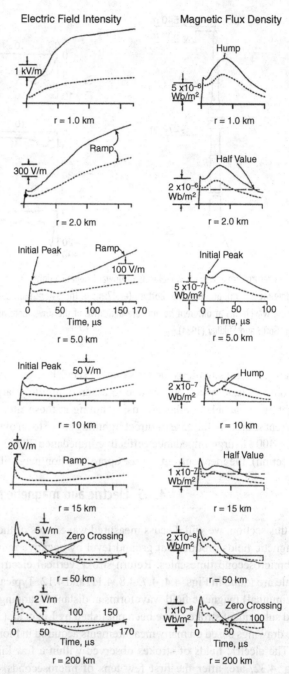

Fig. 4.32   Typical vertical electric field intensity (left column) and azimuthal magnetic flux density (right column) waveforms for first (solid line) and subsequent (broken line) return strokes at distances of 1, 2, 5, 10, 15, 50, and 200 km. Adapted from Lin et al. (1979).

fields have essentially identical waveshapes and are usually bipolar, as illustrated in Fig. 4.32. The data of Lin et al. (1979) suggest that at a distance of 50 km and beyond, both electric and magnetic field waveshapes are dominated by their respective radiation components.

The initial field peak evident in the waveforms of Fig. 4.32 is the dominant feature of the electric and magnetic field waveforms beyond about 10 km; this initial peak also is a significant feature of waveforms from strokes between a few and about 10 km and can be identified, with some effort, in waveforms for strokes as close as a kilometer. The initial field peak is due to the radiation component of the total field and, hence, as discussed in Chapter 5, decreases inversely with distance in the absence of significant propagation effects. The field peaks produced by different return strokes at known distances can be range normalized for comparison, for example, to 100 km by multiplying the measured field peaks by $r/10^5$, where r is the stroke distance in meters. The geometric mean of the electric field initial peak value, normalized to 100 km, is typically about 6 V m^{-1} for first strokes and 3 V m^{-1} for subsequent strokes. Since the initial electric field peak appears to obey a log-normal distribution, the geometric mean value (equal to the median value for a log-normal distribution) is probably a better characteristic of the statistical distribution of this parameter than the mean (arithmetic mean) value. Note that the geometric mean value for a log-normal distribution is lower than the corresponding mean value and higher than the modal (most probable) value.

Details of the shape of the return-stroke field rise to peak and the fine structure after the initial peak are shown in Fig. 4.33. Note that also shown in Fig. 4.33 are stepped-leader and dart-stepped-leader pulses occurring prior to the return stroke pulse. As illustrated in Fig. 4.33, first-return-stroke field waveforms have a "slow front" (below the dotted line in Fig. 4.33 and labeled F) that rises in a few microseconds to an appreciable fraction of the field peak. Weidman and Krider (1978) found a mean slow front duration of about 4.0 μs with 40 to 50 percent of the initial field peak attributable to the slow front. The slow front is followed by a "fast transition" (labeled R in Fig. 4.33) to peak with a 10–90 percent rise time of about 0.1 μs when the field propagation path is over salt water. As illustrated in Fig. 4.33, fields from subsequent strokes have fast transitions similar to those of first strokes except that these transitions account for most of the rise to peak, while the slow fronts are of shorter duration than for first strokes, typically 0.5–1 μs and only comprise about 20 percent of the total rise to peak (Weidman and Krider 1978).

The higher-frequency content of the return-stroke fields is preferentially degraded in propagating over a finitely conducting earth, so that the fast transition and other rapidly changing fields can only be adequately observed if the propagation path from the lightning to the antenna is over salt water, a relatively good conductor.

4.5.6 Peak currents inferred from measured fields

Lightning peak currents can be estimated from measured electric or magnetic fields, for which a field-to-current conversion procedure (a model-based or empirical formula) is required. The ground-level vertical component of electric field and the azimuthal component of magnetic field are usually employed.

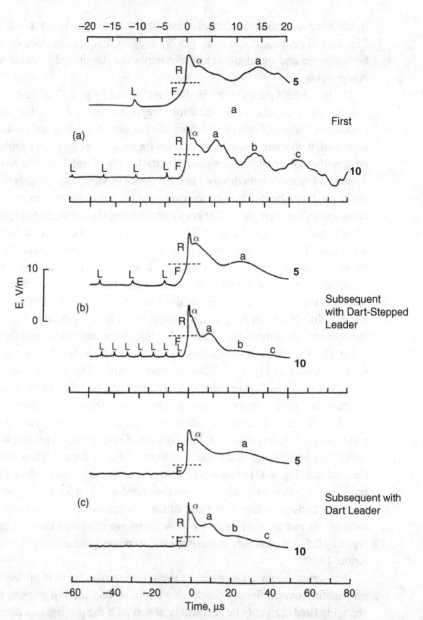

Fig. 4.33 Electric field waveforms of (a) a first return stroke, (b) a subsequent return stroke initiated by a dart-stepped leader, and (c) a subsequent return stroke initiated by a dart leader, showing the fine structure both before and after the initial field peak. Each waveform is shown on two timescales: 5 μs per division (labeled 5) and 10 μs per division (labeled 10). The fields are normalized to a distance of 100 km. L denotes individual leader pulses, F slow front, and R fast transition. Also marked are the small secondary peak or shoulder a and the larger subsidiary peaks a, b, and c. Adapted from Weidman and Krider (1978).

Rakov et al. (1992) proposed the following empirical formula (linear regression equation) to estimate the negative return-stroke peak current, I, from the initial electric field peak, E, and distance, r, to the lightning channel:

$$I = 1.5 - 0.037Er \tag{4.11}$$

where I is in kA and taken as negative, E is positive and in V/m, and r is in km. Equation 4.11 was derived using data for 28 triggered-lightning strokes acquired by Willett et al. (1989) at the Kennedy Space Center (KSC), Florida. The fields were measured at about 5 km and their initial peaks were assumed to be pure radiation. The currents were directly measured at the lightning channel base.

Lightning peak currents can also be estimated using the radiation-field-to-current conversion equation based on the transmission line (TL) model (Uman and McLain, 1969), which for the electric field is given by

$$I = \frac{2\pi\varepsilon_0 c^2 r}{v} E \tag{4.12}$$

where ε_0 is the permittivity of free space, c is the speed of light, and v is the return-stroke speed (assumed to be constant). The return-stroke speed is generally unknown and its range of variation is from one-third to two-thirds of the speed of light. Both I and E in Equation 4.12 are absolute values. The equation is thought to be valid for instantaneous values of E and I at early times (for the initial rising portion of the waveforms, including the peak).

The US National Lightning Detection Network (NLDN) uses an empirical formula, based on triggered-lightning data, to estimate the return-stroke peak current from the measured magnetic field peak and distance to the strike point determined using data from multiple sensors. The conversion procedure includes compensation for the field attenuation due to its propagation over lossy ground (Cummins and Murphy, 2009). The procedure has been validated using rocket-triggered lightning data and tower-initiated lightning data. Since strokes in triggered and tower-initiated lightning are similar to subsequent strokes in natural lightning, the results are applicable only to subsequent strokes. Peak-current estimation errors for negative first strokes and for positive lightning are presently unknown.

Mallick et al. (2013b) compared peak currents obtained using the three methods outlined above with directly measured peak currents for 91 negative strokes in 24 lightning flashes triggered using the rocket-and-wire technique at Camp Blanding, Florida, in 2008–2010. The empirical formula (Eq. 4.11), based on data from the KSC, tended to overestimate peak currents, whereas the NLDN-reported peak currents were on average underestimates. The field-to-current conversion equation based on the transmission line model gave the best match with directly measured peak currents for return-stroke speeds between $c/2$ and $2c/3$ (1.5 and 2×10^8 m/s, respectively).

NLDN-reported peak currents vs. directly measured currents for negative strokes in rocket-triggered lightning are shown in Fig. 4.34. Median absolute current estimation error is 14 percent. Similar scatter plots were reported for peak currents estimated by the NLDN-type European Cooperation for Lightning Detection (EUCLID) network vs. those directly measured at the Gaisberg Tower in Austria.

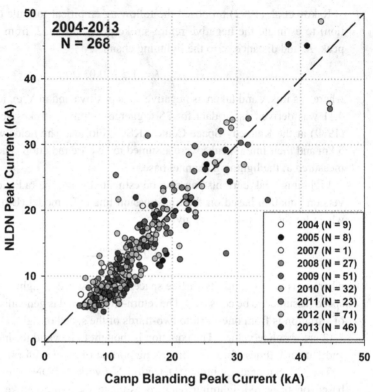

Fig. 4.34 NLDN-reported peak current versus peak current directly measured for 268 return strokes in lightning flashes triggered at Camp Blanding, Florida, in 2004–13. The slanted broken line (slope = 1) is the locus of the points for which the NLDN-reported peak currents and the directly measured peak currents are equal. The NLDN tends to underestimate the peak current (by about 10 percent, on average). Adapted from Mallick et al. (2014).

4.6 Subsequent leader

Return strokes subsequent to the first are usually initiated by dart leaders. An example of a dart leader/return stroke sequence imaged by a streak camera is found in Fig. 4.28a. The negatively sloped image on the left is due to the dart leader, and the brighter image on the right is due to the return stroke. As seen in Fig. 4.28a, the dart leader appears, unlike the stepped leader (see Fig. 4.9), to move continuously, that is, the lowest portion of the leader channel, called the dart, remains luminous during the channel extension from the cloud to ground. Luminosity versus time at different heights above ground corresponding to the streak photograph in Fig. 4.28a is shown for the entire visible channel and for the bottom 480 m in Figs. 4.28b and c, respectively. Many subsequent return strokes (more than one-third of second strokes in channels that are not too old) are initiated by leaders that exhibit pronounced stepping in the bottom portion of the channel. Such leaders produce regular pulse sequences that are observed just prior to the return-stroke pulse in distant electric or

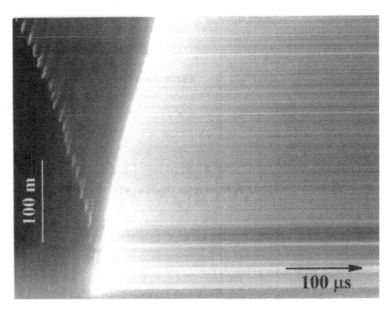

Fig. 4.35 Streak-camera record of a dart-stepped leader/return stroke sequence in negative rocket-triggered lightning (Camp Blanding flash 9734). Time advances from left to right. Courtesy of V. P. Idone, University at Albany, SUNY.

magnetic field records and are called dart-stepped leaders. A streak photograph of a dart-stepped leader in a triggered lightning flash is presented in Fig. 4.35. Some subsequent strokes are initiated by "chaotic leaders," which are identified, in distant electric or magnetic field records, by irregular pulse sequences just prior to the return stroke. When a subsequent leader deflects from the previously formed channel, it continues as a stepped leader, with more than one-third of second-stroke leaders exhibiting this behavior. Subsequent strokes creating a new termination on the ground are intermediate in many of their characteristics between first strokes initiated by stepped leaders and subsequent strokes following a previously formed channel. It is common practice to term all the leaders preceding subsequent return strokes in previously formed channels "dart leaders."

The typical duration of the subsequent leader following a previously formed channel is 1 to 2 ms. For subsequent strokes creating a new termination on ground, it can be considerably longer. The dart leader progresses downward at a typical speed of 10^7 m s^{-1} and deposits a total charge along the channel of the order of 1 C. The dart-leader peak current is about 1 kA. The formation of each step of the dart-stepped leader is associated with a charge of a few millicoulombs and a current of a few kiloamperes.

Examples of the overall electric field changes of subsequent leaders together with the corresponding return-stroke field changes are found in Fig. 4.12 (left panel, the bottom set of waveforms) and Fig. 4.36. The stroke presented in Fig. 4.36 was the fifth stroke in a flash. That stroke followed a previously formed channel. The subsequent stroke presented in Fig. 4.12 was the second stroke in a flash, which created a new termination on ground. Figure 4.12 also presents a VHF image of the subsequent leader channel (right panel, the

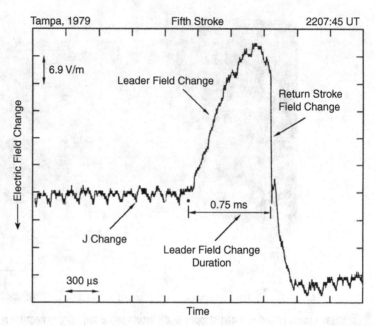

Fig. 4.36   Electric field change of the fifth stroke in a five-stroke flash which occurred in Florida at a distance of 7.6 km. A positive field change (atmospheric electricity sign convention, according to which a downward-directed electric field or field change is defined as positive) deflects downward. The small solid circle near the middle of the figure marks the starting point of the leader field change. The time interval labeled 0.75 ms indicates the leader field change duration. Note that the leader electric field change is hook-shaped. Adapted from Rakov and Uman (1990c).

bottom image labeled L2). Electric field waveforms produced by dart leaders in triggered lightning recorded at distances ranging from 30 to 110 m are shown in Fig. 4.37. Similar to negative-first-stroke leaders, the shape of electric field waveforms produced by negative-subsequent-stroke leaders changes from hook-like net negative at close distances to mono-tonic positive at far distances (the polarity here corresponds to the atmospheric electricity sign convention, according to which a downward-directed electric field change vector is assumed to be positive).

4.7 Continuing current

Continuing current is usually defined as a relatively low-level current of typically tens to hundreds of amperes that immediately follows a return stroke, in the same channel to the ground, and typically lasts for tens to hundreds of milliseconds. This current can be viewed as a quasi-stationary arc between the cloud charge source and ground along the path created by the preceding leader/return stroke sequence or sequences. Relatively short perturbations in the continuing current, typically lasting for a few milliseconds or less, are called M

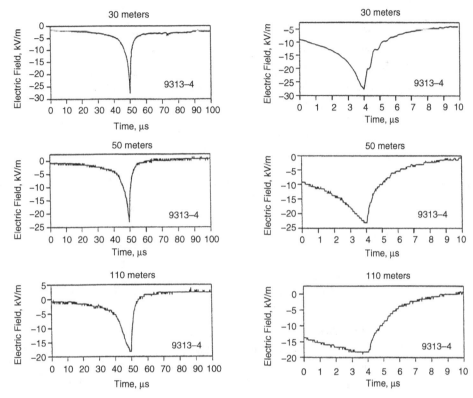

Fig. 4.37 Electric field waveforms of the fourth leader/return stroke sequence of the five-stroke flash 9313 as recorded at 30, 50, and 110 m at Camp Blanding, Florida, shown on 100 μs and 10 μs timescales (left-hand and right-hand columns, respectively). The initial downward-going portion of the waveform is due to the dart leader, and the upward-going portion is due to the return stroke. Adapted from Rakov et al. (1998).

components (Section 4.8). Due to their relatively large charge transfers, continuing currents are responsible for most serious lightning damage associated with thermal effects, such as burned holes in the metal skins of aircraft, burned-through ground wires of overhead power lines, and forest fires.

Most of the published information on continuing currents in negative lightning, particularly information regarding the occurrence of this lightning process, concerns only "long" continuing currents. The latter have been arbitrarily defined as currents to ground lasting in excess of 40 ms. In this section, if not specified otherwise, the term continuing current means "long" (longer than 40 ms) continuing current.

The distinction between return-stroke current and continuing current is apparently related to the source of the charge that is transported to ground by these two lightning processes. The return stroke primarily removes charge that has been deposited on the channel by a preceding leader, whereas continuing current is likely associated with the

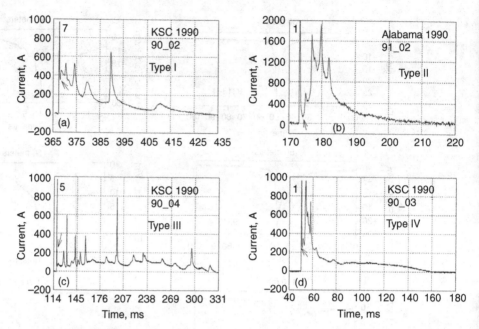

Fig. 4.38 Examples of continuing current waveshapes in negative triggered lightning. In each figure, the arrow indicates the assumed beginning of the continuing current. (a) Type I, more or less exponential decay with superimposed M-current pulses; (b) Type II, a hump with superimposed M-current pulses followed by a relatively smooth decay; (c) Type III, a slow increase and decrease of current with superimposed M-current pulses throughout; (d) Type IV, a hump with superimposed M-current pulses followed by a steady plateau without pronounced pulse activity. The number in the upper left corner indicates the order of the return stroke in the flash (triggered-lightning strokes are similar to subsequent strokes in natural lightning). The return-stroke current pulses are clipped. Adapted from Fisher et al. (1993).

tapping of fresh charge regions in the cloud. It is generally thought that the maximum duration of the return-stroke stage is about 3 ms.

The characteristic electric field signature of continuing current (labeled C Field Change) is seen in Fig. 4.4, the top set of records. It is a large slow electric field change of the same polarity as that of the preceding return stroke. The overwhelming majority of long continuing currents are initiated by the subsequent strokes of multiple-stroke flashes as opposed to either the first stroke in a multiple-stroke flash or the only stroke in a single-stroke flash.

Different types of continuing current waveshapes in rocket-triggered lightning observed by Fisher et al. (1993) are shown in Fig. 4.38. Waveshapes of Types I and II were most common. Continuing currents often appear to begin with a current pulse characteristic of the M component, on average 1.4 ms after the return stroke initial current peak.

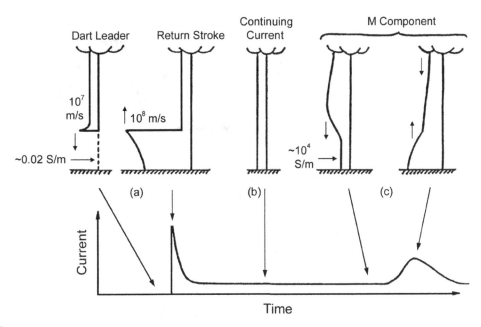

Fig. 4.39 Schematic representation of current versus height profiles for three modes of charge transfer to ground in negative lightning subsequent strokes: (a) dart leader/return stroke sequence, (b) continuing current, and (c) M-component. The corresponding current versus time waveform represents the current at the ground. © Vladimir A. Rakov and Martin A. Uman 2003, published by Cambridge University Press, reprinted with permission.

4.8 M-component

M components are perturbations (or surges) in the relatively steady continuing current and in the associated channel luminosity. The M in the term M component stands for D. J. Malan, who was the first to study this lightning process. "Classical" M-components occur in downward lightning and involve a single channel between the cloud base and the ground. They are excited at the upper extremity of the channel (in the cloud) by either recoil leaders or separate in-cloud leaders coming in contact with the grounded current-carrying channel. Recoil leaders are thought to occur in decayed channel branches previously supplying current to the grounded channel. Similar to leader/return stroke sequences and to continuing currents in negative lightning, M components serve to transport negative electric charge from the cloud to the ground. The M-component mode of charge transfer to ground differs from the dart leader/return stroke mode in that the former requires the presence of a current-carrying channel to the ground, while the latter apparently occurs along the remnants of the previously formed channel when there is essentially no current (above 0.1–2 A) flowing to the ground. Different modes of charge transfer to the ground are illustrated in Fig. 4.39. M-component-like processes also occur during the initial stage of object-initiated and

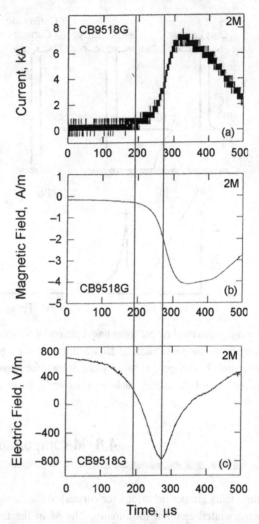

Fig. 4.40 (a) Channel-base current (b) magnetic field, and (c) electric field for a large M component that followed the second stroke of a rocket-triggered lightning flash at Camp Blanding, Florida. The fields were recorded at a distance of 280 m from the lighting channel. The left vertical line indicates the position of current pulse onset and the right vertical line the position of negative electric field peak. Note that the onset of the M-component current pulse at the channel base occurs some tens of microseconds before the electric field peak. Adapted from Rakov et al. (1998).

rocket-triggered lightning, when a steady-current-carrying channel is excited at a height of a kilometer or more above its ground termination point.

M-component current pulses recorded at the lightning channel base are seen in Fig. 4.38. M-component currents typically have peaks of a few hundreds of amperes, rise times of tens to hundreds of microseconds, and charge transfers of a few hundreds of millicoulombs. Some M components have current peaks in the kiloamperes range (see, for example,

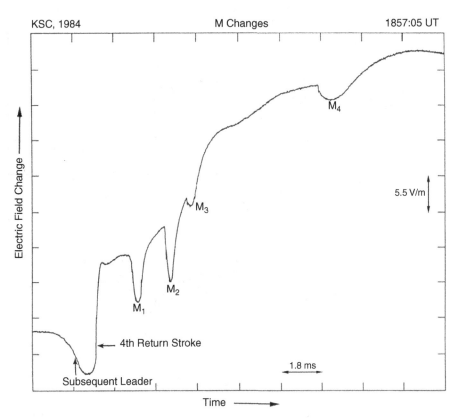

KSC, 1984   M Changes   1857:05 UT

Fig. 4.41 Hook-shaped electric field changes due to M components (labeled M_1 through M_4) during the continuing current initiated by the fourth return stroke in a flash that occurred in Florida at a range of probably 5 km. A positive field change (atmospheric electricity sign convention, according to which a downward-directed electric field or field change is defined as positive) deflects upward. Adapted from Rakov et al. (1992a).

Fig. 4.40a), comparable to current peaks of smaller return strokes. An electric field record showing signatures of M-components is presented in Fig. 4.41 (see also electric field signatures labeled "M-change" and corresponding optical images in Fig. 4.4, the upper set of records).

Rakov et al. (1995) have proposed a mechanism for the "classical" M-component. According to this mechanism, an M component is essentially a guided-wave process that involves a downward progressing incident wave (the analog of a leader) and an upward progressing wave that is a reflection of the incident wave from the ground (the analog of a return stroke). Ground is sensed by the incident M-wave as essentially a short circuit, so the reflection coefficient for current at ground is close to +1, and the reflection coefficient for the associated charge density is close to −1. Because the reflection coefficients for the traveling waves of current and charge density are positive and negative, respectively, the incident and reflected waves of current (which determine the close magnetic fields) add at each channel section, while the incident and reflected waves of charge density (which

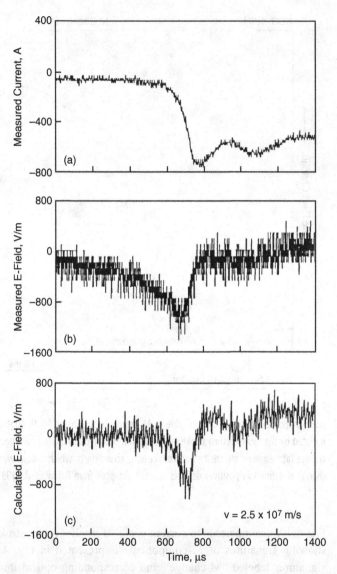

Fig. 4.42 (a) Measured channel-base current, (b) measured 30 m electric field, and (c) calculated 30 m electric field for M component that occurred about 2 ms after the last stroke of five-stroke flash 9313 triggered at Camp Blanding, Florida. The value of current-wave speed (adjustable model parameter) providing the best field-magnitude match with the measurement in (b) was 2.5×10^7 m/s. Adapted from Rakov et al. (1995).

determine the close electric fields) subtract. As a result, at close ranges, the M-component magnetic field has an overall waveshape similar to that of the channel-base current, whereas the M-component electric field has a waveform that appears to be the time derivative of the channel base current, as seen in Fig. 4.40.

Rakov et al. (1995) tested the validity of a simple model based on the guided-wave mechanism using (1) measured channel-base M-component current, (2) electric field measured 30 m from the channel, and (3) adjustable speed v, which essentially controls the M electric field magnitude and has relatively little effect on the field waveshape. The model-predicted electric field 30 m from the channel for one M component, along with measured field and current and with the value of v providing the best field magnitude match with measurements, is shown in Fig. 4.42. There is a fairly good agreement between the calculated and observed electric fields. In particular, the model predicted electric field peak occurs appreciably later than the onset of the M-component current pulse, as does the measured field peak. This is a distinctive feature (see also Fig. 4.40) of "classical" M-components, which indicates that the M-component current pulse at the channel base begins while the bulk of the associated charge is still being transported toward the ground in the upper channel sections, as if the return-stroke-like process was commencing at ground before the arrival of leader-like process there.

Recently, Jiang et al. (2013) proposed an M-component model based on the guided-wave concept of this process described above. In their model, the upward wave serves to deplete the negative charge of the downward wave at their interface and makes the charge density below the interface to be approximately zero.

4.9 J- and K-processes

The J- or "junction"-process takes place in the cloud during the time interval between return strokes. It is identified by a relatively steady electric field change on a timescale of tens of milliseconds. The J-change can be of either the same or opposite polarity as that of the return-stroke electric field change. In the former case, it is generally smaller than the field change due to continuing current (Section 4.7), and, as differentiated from the continuing current field change, is not associated with a luminous channel between cloud and ground. Relatively rapid electric field variations termed K-changes also occur between strokes, generally at intervals ranging from some milliseconds to some tens of milliseconds and, hence, appear to be superimposed on the overall electric field change associated with the J-process. The electric field change following the last stroke (after the end of continuing current, if any) is probably of the same nature as the J-change discussed above. Both J-processes and K-processes in cloud-to-ground discharges serve to effectively transport fresh negative charges into and along the existing channel (or its remnants), although not all the way to ground. It is likely that most of the K processes in a ground flash can be viewed as unsuccessful or attempted subsequent leaders.

In electric field records not influenced by instrumental decay (Chapter 7), K-changes are usually identified as the step-like (or ramp-like) electric field changes that occur during inter-stroke intervals and after the last stroke, have the same polarity as the J change, and have a 10–90 percent rise time of 3 ms or less. Examples of such K-changes are seen in Fig. 4.43. Similar K-changes occur in cloud discharges. K-change signatures (distorted by the fast instrumental decay), labeled "K-Change", are also seen in Fig. 4.4, the upper set of records).

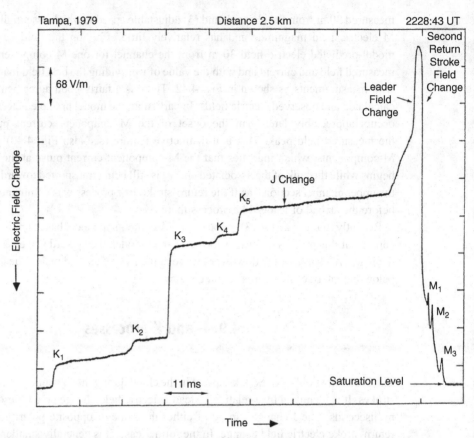

Fig. 4.43 A portion of the electric field record for a flash that occurred in Florida at a distance of 2.5 km. Labeled are five pronounced K changes (K_1 through K_5), J change, leader and return-stroke field changes, and three field changes due to M components (M_1 through M_3). A positive field change (atmospheric electricity sign convention, according to which a downward-directed electric field or field change is defined as positive) deflects downward. Adapted from Thottappillil et al. (1990).

4.10 Summary

Each lightning stroke is composed of a downward-moving leader and an upward-moving return stroke. The leader creates a conductive path between the cloud charge source and ground and deposits negative charge along this path, while the return stroke traverses that path moving from ground toward the cloud charge source region and neutralizes the negative leader charge. The stepped leader initiating the first return stroke moves intermittently, while subsequent-stroke leaders usually appear to move continuously.

Following the initial breakdown, possibly between the main negative and lower positive charge regions in the cloud, the stepped leader propagates toward the ground at an average speed of 2×10^5 m s^{-1}. Below the cloud lower boundary, each leader step is typically 1 μs in duration and tens of meters in length, with a time interval between steps of 20 to 50 μs. The average stepped-leader current is some hundreds of amperes, and the peak value of the current pulse associated with an individual step is at least 1 kA. The transition from the leader stage to the return-stroke stage is referred to as the attachment process. The upward propagation speed of a return stroke below the cloud lower boundary is typically between one-third and one-half of the speed of light: that is, about three orders of magnitude higher than the stepped-leader speed. The first return stroke current measured at ground rises to an initial peak typically of about 30 kA (median value) in some microseconds and decays to a half-peak value in some tens of microseconds. The return stroke lowers to ground the several coulombs of charge originally deposited on the stepped-leader channel.

Subsequent strokes occur after the essential cessation of current flow to ground. In-cloud processes termed J-processes involve a redistribution of cloud charges on a tens of milliseconds timescale in response to the preceding return stroke. Transients occurring during the slower J-process are referred to as K-processes. Both J-processes and K-processes in cloud-to-ground discharges serve to effectively transport fresh negative charges into and along the existing channel (or its remnants), although not all the way to ground. The dart leader progresses downward at a typical speed of 10^7 m s^{-1} and deposits a total charge along the channel of the order of 1 C. The dart-leader peak current is about 1 kA. Subsequent return-stroke current measured at ground rises to a peak value of 10 to 15 kA in less than a microsecond and decays to half-peak value in a few tens of microseconds. The maximum current derivative is typically of the order of 100 kA μs^{-1}. The average upward propagation speed of subsequent return strokes is similar to that of first return strokes. The impulsive, kiloampere-scale component of a subsequent stroke current is often followed by a continuing current which has a magnitude of tens to hundreds of amperes and a duration up to hundreds of milliseconds. Transient processes that occur during the continuing-current stage and serve to transport negative charge to ground are referred to as M components.

Questions and problems

4.1. Sketch the total electric field change vs. time due to a negative, uniformly charged lightning leader developing vertically toward ground from a spherically symmetrical cloud charge region. On the same sketch, show the field change components due to the developing leader channel and due to the depletion of the cloud charge. Assume that the height of the cloud charge center above ground is 5 km, the line charge density on the leader channel is 10^{-3} C/m, the leader extension speed is 10^6 m/s, and the horizontal distance from the cloud charge center to the field point is (a) 1 km, (b) 5 km, and (c) 15 km.

4.2. Prove that the ratio of the net electric field change at the ground due to the leader, to that due to the return stroke (the leader-to-return-stroke electric field change ratio), approaches -1 as the distance approaches zero. Explain why. Assume that the leader channel is vertical and uniformly charged and that the return stroke neutralizes all the leader charge and deposits no extra charge onto the leader channel.

4.3. Prove that the leader-to-return-stroke electric field change ratio is equal to zero when the elevation angle of the cloud charge source (the angle between the line joining the charge source and field point and ground surface) is about 52°. Assume that the leader channel is vertical and uniformly charged and that the return stroke neutralizes all the leader charge and deposits no extra charge onto the leader channel.

4.4. At far ranges, the leader-to-return-stroke electric field change ratio is close to +1, which is consistent with a uniform charge density distribution along the leader channel. How would the ratio change if the line charge density distribution were strongly skewed toward the bottom of the channel?

4.5. Sketch the total electric field change vs. time due to a typical subsequent leader-return stroke sequence lowering negative charge to earth at (a) 2 km, (b) 20 km, and (c) 200 km. Use as many timescales at each distance as necessary to show all significant waveform characteristics. Why does the field-change waveshape vary with distance?

4.6. Draw the total flash electric filed change vs. time for a typical three-stroke flash lowering negative charge to earth at (a) 1 km, (b) 10 km, and (c) 100 km. Use as many timescales at each distance as necessary to show all significant waveform characteristics.

4.7. Compare the characteristics of first and subsequent strokes in negative CGs. Discuss the differences.

4.8. Why is the return-stroke speed lower than the speed of light?

Further reading

Bazelyan, E. M. and Raizer, Yu. P. (2000a). *Lightning Physics and Lightning Protection*, 325 pp., Bristol: IOP Publishing.

Cooray, V., ed. (2014). *The Lightning Flash*, 2nd Edition, 896 pp., London: IET.

CIGRE WG C4.407 Technical Brochure 549, Lightning Parameters for Engineering Applications, V. A. Rakov, Convenor (US), A. Borghetti, Secretary (IT), C. Bouquegneau (BE), W. A. Chisholm (CA), V. Cooray (SE), K. Cummins (US), G. Diendorfer (AT), F. Heidler (DE), A. Hussein (CA), M. Ishii (JP), C. A. Nucci (IT), A. Piantini (BR), O. Pinto, Jr. (BR), X. Qie (CN), F. Rachidi (CH), M. M. F. Saba (BR), T. Shindo (JP), W. Schulz (AT), R. Thottappillil (SE), S. Visacro (BR), W. Zischank (DE), 117 pp., August 2013.

Rakov, V. A. and Uman, M. A. (2003). *Lightning: Physics and Effects*, 687 pp., New York: Cambridge.

5 Calculation of lightning electromagnetic fields

5.1 General approach

The electric field intensity, $\overline{E}$, and magnetic flux density, $\overline{B}$, are usually found using the scalar, ϕ, and vector, $\overline{A}$, potentials as follows:

$$\overline{E} = -\nabla\phi - \partial\overline{A}/\partial t \qquad (5.1)$$

$$\overline{B} = \nabla \times \overline{A} \qquad (5.2)$$

The potentials, ϕ and $\overline{A}$, are related to the volume charge density, ρ, and current density, $\overline{J}$, respectively, by

$$\phi = \frac{1}{4\pi\varepsilon_0} \int_{V'} \frac{\rho(r', t - R/c)}{R} dV' \qquad (5.3)$$

$$\overline{A} = \frac{\mu_0}{4\pi} \int_{V'} \frac{\overline{J}(r', t - R/c)}{R} dV' \qquad (5.4)$$

where ε_0 and μ_0 are the electric permittivity and magnetic permeability of free space (also applicable to air), respectively, V' is the source volume, dV' is the differential source volume, $\overline{r}'$ and $\overline{r}$ are the radius (position) vectors of the differential source volume and observation (field) point, respectively, t is the time, c is the speed of light, and R is the inclined distance between the differential source volume and the observation point. The ratio R/c is the time required for electromagnetic signals to propagate from the source to the field point, and t – R/c is referred to as the retarded time. Note that $\overline{E}$, $\overline{B}$, $\overline{A}$, and $\overline{J}$ are vectors and ϕ and ρ are scalars.

The scalar and vector potentials are related by the Lorentz condition:

$$\nabla \cdot \overline{A} + c^{-2}\partial\phi/\partial t = 0 \qquad (5.5)$$

which is equivalent to the continuity equation relating ρ and $\overline{J}$:

$$\nabla \cdot \overline{J} + \partial\rho/\partial t = 0 \qquad (5.6)$$

It follows that the source quantities, ρ and $\bar{J}$, cannot be specified independently; they must satisfy the continuity equation (Eq. 5.6). Alternatively, one can specify $\bar{J}$, find $\bar{A}$ from Eq. 5.4, and use Eq. 5.5 to find ϕ. In this latter case, there is no need to specify ρ in computing electric fields.

5.2 Field equations

The most general equations for computing the vertical electric field E_z and azimuthal magnetic field B_ϕ due to an upward-moving lightning return stroke for the case of a field point P on perfectly conducting ground (see Fig. 5.1) and a return-stroke model (see Chapter 6) without current discontinuity at the moving front are given below:

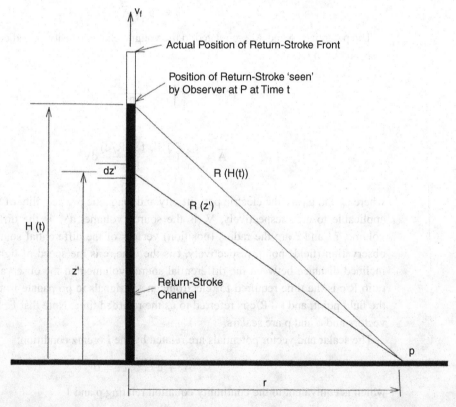

Fig. 5.1 Geometry used in deriving equations for electric and magnetic fields at point P on the ground (assumed to be perfectly conducting) a horizontal distance r from the vertical lightning return-stroke channel extending upward with speed v_f. Adapted from Thottappillil et al. (1997).

$$E_z(r, t) = \frac{1}{2\pi\varepsilon_0} \int_0^{H(t)} \left[\frac{2z'^2 - r^2}{R^5(z')} \int_{\frac{z'}{v_f} + \frac{R(z')}{c}}^t I\left(z', \tau - \frac{R(z')}{c}\right) d\tau \right.$$

$$\left. + \frac{2z'^2 - r^2}{cR^4(z')} I\left(z', t - \frac{R(z')}{c}\right) - \frac{r^2}{c^2R^3(z')} \frac{\partial I(z', t - R(z')/c)}{\partial t} \right] dz' \qquad (5.7)$$

$$B_\phi(r, t) = \frac{\mu_0}{2\pi} \int_0^{H(t)} \left[\frac{r}{R^3(z')} I\left(z', t - \frac{R(z')}{c}\right) + \frac{r}{cR^2(z')} \frac{\partial I(z', t - R(z')/c)}{\partial t} \right] dz' \qquad (5.8)$$

where v_f is the propagation speed of the return-stroke front and $H(t)$ is the height of the return-stroke front as "seen" by the observer at time t (see Fig. 5.1). This height can be found from the following equation:

$$t = \frac{H(t)}{v_f} + \frac{R(H(t))}{c} \qquad (5.9)$$

Derivation of Eqs. 5.7 and 5.8 is given in Appendix 3.

A lightning return-stroke model is needed to specify $I(z',t)$. Equations 5.7 and 5.8 are suitable for computing fields at ground level using the electromagnetic, distributed-circuit, or "engineering" return-stroke models to be discussed in Chapter 6. Some of the "engineering" models include a current discontinuity at the moving front. Such a discontinuity is an inherent feature of some traveling-current-source-type models (BG and TCS), even when the current at the channel base starts from zero. The transmission-line-type models may include a discontinuity at the front if the channel-base current starts from a nonzero value. The DU model does not allow a current discontinuity either at the upward-moving front or at the channel base. The three terms in Eq. 5.7 are referred to as the electrostatic, induction (or intermediate), and electric radiation field components, respectively, and the two terms in Eq. 5.8 are referred to as the magnetostatic (or induction) and magnetic radiation field components, respectively. For return-stroke models with current discontinuity at the moving front, Eqs. 5.7 and 5.8 describe the fields only due to sources below the upward-moving front. A current discontinuity at the moving front gives rise to an additional term in each of Eqs. 5.7 and 5.8:

$$E_z^{disc} = -\frac{1}{2\pi\varepsilon_0} \frac{r^2}{c^2R^3(H(t))} I\left(H(t), \frac{H(t)}{v_f}\right) \frac{dH(t)}{dt} \qquad (5.10)$$

$$B_\phi^{disc} = \frac{\mu_0}{2\pi} \frac{r}{cR^2(H(t))} I\left(H(t), \frac{H(t)}{v_f}\right) \frac{dH(t)}{dt} \qquad (5.11)$$

Note that the front discontinuity produces only a radiation field component, no electrostatic or induction field components.

We now use the transmission-line (TL) model (Chapter 6) to derive the far-field approximation to Eq. 5.7. At far distances (typically at $r \geq 50$ km), the radiation field component is dominant, so that we can write:

$$E_z(t) \approx E_z^{rad}(t) = -\frac{1}{2\pi\varepsilon_0}\int_0^{H(t)} \frac{r^2}{c^2R^3} \frac{\partial I(z', t - R/c)}{\partial t}dz' \tag{5.12}$$

From Eq. 5.9, noting that for the TL model $v_f = v$ and that at far ranges $R \approx r$, we have

$$H(t) = v(t - R/c) \approx v(t - r/c) \tag{5.13}$$

Thus, Eq. 5.12 becomes

$$E_z^{rad}(t) = -\frac{1}{2\pi\varepsilon_0 c^2 r}\int_0^{v(t-r/c)} \frac{\partial I(z', t - r/c)}{\partial t}dz' \tag{5.14}$$

For a current wave traveling at constant speed v in the positive z′ direction, $I(t-z'/v)$, the time derivative of current can be converted to the spatial derivative by comparing the following two equations, in which the chain rule was used to take the partial derivatives with respect to z′ and t and I′ stands for the derivative of I with respect to $(t - z'/v)$:

$$\frac{\partial I(t - z'/v)}{\partial z'} = \frac{\partial I(t - z'/v)}{\partial(t - z'/v)}\frac{\partial(t - z'/v)}{\partial z'} = -\frac{1}{v}I'(t - z'/v)$$

$$\frac{\partial I(t - z'/v)}{\partial t} = \frac{\partial I(t - z'/v)}{\partial(t - z'/v)}\frac{\partial(t - z'/v)}{\partial t} = I'(t - z'/v)$$

from which

$$\frac{\partial I(t - z'/v)}{\partial t} = -v\frac{\partial I(t - z'/v)}{\partial z'}$$

For the TL model in terms of retarded time, $I(z', t - r/c) = I(0, t - z'/v - r/c)$ and

$$\frac{\partial I(0, t - z'/v - r/c)}{\partial t} = -v\frac{\partial I(0, t - z'/v - r/c)}{\partial z'}$$

so that Eq. 5.14 becomes

$$E_z^{rad}(t) = \frac{v}{2\pi\varepsilon_0 c^2 r}\int_{z'=0}^{z'=v(t-r/c)} dI(0, t - z'/v - r/c)$$

$$= \frac{v}{2\pi\varepsilon_0 c^2 r}\left[I\left(0, t - \frac{v(t - r/c)}{v} - r/c\right) - I(0, t - r/c)\right]$$

$$= \frac{v}{2\pi\varepsilon_0 c^2 r}[I(0, 0) - I(0, t - r/c)]$$

Normally $I(0, 0) = 0$, so that

$$E_z^{rad}(t) = -\frac{v}{2\pi\varepsilon_0 c^2 r}I(0, t - r/c) \tag{5.15}$$

that is, the electric radiation field component is proportional to the channel-base (z′=0) current. The minus sign indicates that for a positive current (positive charge moving upward) the electric field vector is directed downward. The corresponding magnetic radiation field can be found from $|B_\phi^{rad}| = |E_z^{rad}|/c$ or $|H_\phi^{rad}| = |E_z^{rad}|/\eta_0$, where $H_\phi = B_\phi/\mu_0$ is the magnetic field intensity and $\eta_0 = 377\ \Omega$ is the intrinsic impedance of free space. Equation 5.15 and its magnetic field counterpart are further discussed in Section 5.5, along with the close field approximations derived using other than radiation field components.

Krider (1994), using the TL model (Chapter 6), computed the peak electric fields (radiation component) in the upper half-space for different values of the return-stroke speed, v, relative to the speed of light, c. He found that the largest fields are radiated at relatively small polar angles, θ, measured from the vertical when the return-stroke speed is very close to the speed of light. For example, if the peak current is 30 kA, v/c = 0.9, and there is no attenuation in the atmosphere or on ground, the peak electric field at 100 km is 21 V m^{-1}, the peak power density is 1.1 W m^{-2}, and the corresponding polar angle is 29°. For small v/c ratios (less than 0.7 or so) the peak field is radiated along the ground ($\theta = 90°$). Similar results were obtained by Rakov and Tuni (2003) who used the modified transmission line model with exponential current decay with height (MTLE; see Chapter 6). On the other hand, when $\theta = 0°$, the electric radiation field component vanishes (a dipole does not radiate along its axis) and, as a result, the electric field at small polar angles can be dominated by its induction component (Lu, 2006).

Thottappillil et al. (1997) derived an electrostatic field equation in terms of line charge density, ρ_L, for a very close observation point, such that r << H(t), and assuming that (1) retardation effects are negligible and (2) return-stroke line charge density does not vary appreciably with height within the channel section significantly contributing to the field at r,

$$E_z(z,\ t) \approx -\ \frac{\rho_L(t)}{2\pi\varepsilon_0 r} \tag{5.16}$$

Equation 5.16 indicates that the electrostatic field produced by a very close return stroke is approximately proportional to the line charge density on the bottom part of the channel. Note that the right-hand side of Eq. 5.16 is the negative of the right-hand side of Eq. 4.5 derived for a very close, uniformly charged and fully developed leader.

5.3 The reversal distance for electrostatic and induction electric field components of a short current element

For an elevated short dipole that is vertical (see Fig. 5.2), the peak radiation field on the one hand and induction (intermediate) and electrostatic field changes on the other hand may have opposite polarities. This follows from Eq. A.3.16 for the electric field due to a differential vertical current element Idz′ located at height z′ above perfectly conducting ground, in which the first (electrostatic) and second (induction) terms contain $(2z'^2 - r^2)$. The horizontal distance r at which the static and induction components on the ground reverse their direction is equivalent to the reversal distance of the electrostatic field from an elevated, finite-length vertical dipole (see Eq. 3.4).

Motion of positive charge upward (or negative charge downward) produces a radiation electric field change (initial peak) directed downward at all distances, as shown in Fig. 5.2 (inset). Conversely, for positive charge moving downward (or negative charge upward) the

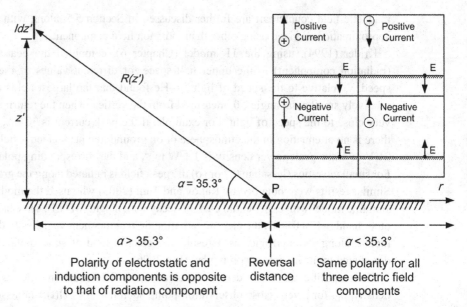

Fig. 5.2 Illustration of the reversal distance for the electrostatic and the induction field components (dipole technique; see Section 5.4). Inset shows the direction of the radiation component of electric field vector for different combinations of the charge polarity and the direction of its motion (also the direction for all three components when α < 35.3°). The direction of the electric field vector refers to the initial half cycle in the case of bipolar waveforms. Adapted from Nag and Rakov (2010).

radiation electric field change is directed upward. At close distances (i.e. for angle α > 35.3° in Fig. 5.2), the motion of positive charge upward produces electrostatic and induction field changes directed upward and the radiation electric field change is directed downward, while at far distances (i.e. for α < 35.3°), all three electric field components are directed downward.

5.4 Non-uniqueness of electric field components

The three components of an electric field (or the change in that field), referred to as the electrostatic, induction (intermediate), and radiation components, are not unique (e.g. Rubinstein and Uman, 1989). For example, these components, often identified by their $1/R^3$, $1/R^2$, and $1/R$ distance dependences, are different for the dipole (Lorentz condition) and monopole (continuity equation) approaches to calculating lightning electric fields. In both approaches, the electric field is found as $\overline{E} = -\nabla\varphi - \partial\overline{A}/\partial t$, where φ is the scalar potential and $\overline{A}$ is the vector potential. The expressions for $\overline{A}$ in the two techniques are the same, but those for φ are different: φ is found from $\overline{A}$ using Lorentz condition in the dipole technique, while in the monopole technique it is found from the charge density, which is related to current via the continuity equation.

Table 5.1 Comparison of electric field components (at ground level) based on dipole (e.g. Uman, 1987) and monopole (Thomson, 1999) techniques for a differential current element Idz' at height z' above ground

Electric field component identified by its distance dependence	Originated from		Polarity	
	Dipole technique	Monopole technique	Dipole technique	Monopole technique
$1/R^3$ (electrostatic)	$\nabla\varphi$	$\nabla\varphi$	Negative for $\alpha < 35.3°$ Positive for $\alpha > 35.3°$	Negative
$1/R^2$ (induction)	$\nabla\varphi$	$\nabla\varphi$	Negative for $\alpha < 35.3°$ Positive for $\alpha > 35.3°$	Negative
$1/R$ (radiation)	$\nabla\varphi$ and $\partial\bar{A}/\partial t$	$\partial\bar{A}/\partial t$	Negative	Negative

φ is the scalar potential (different in the two techniques) and $\bar{A}$ is the vector potential; $\bar{E} = -\nabla\varphi - \partial\bar{A}/\partial t$; $\alpha = \sin^{-1}(z'/R(z'))$.

The two approaches are equivalent – that is, produce identical total fields – although the individual electric field components may even have different polarities, as seen in Table 5.1. Note that, in contrast with the more common dipole (Lorentz condition) technique, there is no reversal distance for the $1/R^3$ and $1/R^2$ components in the monopole technique, and that the $1/R$ component originates only from the time-derivative of magnetic vector potential. (There is another version of the monopole technique (Thottappillil and Rakov, 2001), which yields different $1/R^3$, $1/R^2$, and $1/R$ components, but identical total fields.) The differences between the field components found using different techniques are considerable at close ranges but become negligible at far ranges.

For the dipole approach and electrically short channel segment, Thottappillil and Rakov (2001) have noted that the distance dependences of electric field components are not exactly $1/R^3$, $1/R^2$, and $1/R$, because of the additional dependence on $\sin^2\theta$, where $\sin\theta = r/R$, with r and $R = f(z')$ being the horizontal and inclined distances, respectively, between the source and field points. Only when $\sin^2\theta \approx 1$ (at relatively large distances, when $R \approx r$), are the distance dependences exactly $1/R^3$, $1/R^2$, and $1/R$.

5.5 Short channel segment vs. total radiating channel length

The electric field components discussed in Sections 5.3 and 5.4 above are defined for a differential current element (an electrically short channel segment), for which the current does not vary along the radiator length. In computing lightning electric fields, the

Table 5.2 Approximate expressions for the electromagnetic fields based on the TL model with an arbitrary return-stroke speed v for near and far ranges, converging to exact expressions as v approaches c. Adapted from Chen et al. (2015)

	Approximate expressions based on the TL model with an arbitrary return-stroke speed v	Exact expressions based on the TL model with v = c
Near ranges	$E_z \approx -\dfrac{1}{2\pi\varepsilon_0 vr}I(0,\ t-r/c)$	
Far ranges	$E_z \approx -\dfrac{v}{2\pi\varepsilon_0 c^2 r}I(0,\ t-r/c)$	$E_z = -\dfrac{1}{2\pi\varepsilon_0 cr}I(0,\ t-r/c)(\text{any range})$
Near ranges	$H_\phi \approx \dfrac{1}{2\pi r}I(0,\ t-r/c)$	
Far ranges	$H_\phi \approx \dfrac{v}{2\pi cr}I(0,\ t-r/c)$	$H_\phi = \dfrac{1}{2\pi r}I(0,\ t-r/c)\ (\text{any range})$

integration over the entire radiating channel (over height z′) must be performed at each instant of time, with the inclined distance R being a function of z′. As a result, the horizontal distance r = const is often used instead of R = f(z′) for evaluating the distance dependence of field components produced by the entire radiating channel.

For the transmission line (TL) model (Uman and MacLain, 1969), the sum of electro-static and induction components of electric field at close ranges and the radiation field component at far ranges each vary approximately as 1/r and each are proportional to the channel-base current (Chen et al., 2015). As the current wave propagation speed v approaches c (speed of light), the approximate close-range equation (derived using only the *electrostatic and induction* field components) and the approximate far-range equation (derived using only the *radiation* field component) converge to the same (exact) equation for the *total* electric field, which is valid for *any distance* from the lightning channel (see Table 5.2, which also contains a similar result for magnetic field). Thus, for the transmission line model with v = c, the field components lose their significance. Indeed, in this case, the total electric field (and the total magnetic field) at any distance is proportional to the channel-base current and varies as 1/r (even at very close ranges), as expected for a spherical transverse electromagnetic (TEM) wave (Thottappillil et al., 2001). The approx-imations presented in Table 5.2 are applicable only to the initial portions of the field waveforms, since the TL model is inadequate at later times (see Chapter 6). Also, their ranges of validity depend, besides v, on the current waveshape. For example, when v = 1.3 × 10⁸ m/s, the far electric field approximation is valid beyond 100 km for the typical first stroke and beyond 50 km for the typical subsequent stroke. The close electric field approximation is valid within 100 m for the typical first stroke and only within 10 m for the typical subsequent stroke. The ranges of validity increase with increasing v.

Sometimes, the entire lightning channel is approximated as an electrically short dipole with I(z′,t) = I(0,t). In this case, the radiation field component is proportional to dI/dt, while for the more realistic transmission line model, I(z′,t) = I(0,t − z′/v), where v is the current wave propagation speed, the radiation field component is proportional to the product of I and v. The difference here is similar to the one between a Hertzian (electrically short) dipole and a traveling-wave antenna, and it has important implications for the estimation of peak currents from range-normalized measured peak fields, as done in modern lightning locating systems (Chapter 8). Since the field-to-current conversion procedures in those systems are usually developed for return strokes, assuming that the current peak is proportional to the field peak, they may yield incorrect results for short cloud discharges, because (leaving aside other differences between cloud and ground discharges) the field peak for electrically short radiators is proportional to the current derivative peak, not to the current peak (Nag et al., 2011).

5.6 Channel-base current equation

For the "engineering" models (Chapter 6) in which a vertical lightning channel and a perfectly conducting ground are assumed, the information on the source required for computing the fields usually includes (1) the channel base current (either measured or assumed from typical measurements) and (2) the upward return-stroke front speed, typically assumed to be constant and in a range from 1×10^8 to 2×10^8 m s^{-1}. The return-stroke current waveform at the channel base is often approximated by the Heidler function (Heidler, 1985):

$$I(0,\ t) = \frac{I_0}{\eta} \frac{(t/\tau_1)^n}{(t/\tau_1)^n + 1} e^{-t/\tau_2} \tag{5.17}$$

where I_0, η, τ_1, n, and τ_2 are constants. This function allows one to change conveniently the current peak, maximum current derivative, and associated electrical charge transfer nearly independently by changing I_0, τ_1, and τ_2, respectively. Equation 5.17 reproduces the observed concave rising portion of a typical current waveform, as opposed to the once more commonly used double-exponential function, which is characterized by an unrealistic convex wavefront with a maximum current derivative at t = 0. Sometimes the sum of two Heidler functions with different parameters or the sum of a Heidler function and a double-exponential function is used to approximate the desired current waveshape. The first and subsequent return-stroke current waveforms recommended by the International Electrotechnical Commission (IEC) Lightning Protection Standard are shown in Figs. 5.3a and b, respectively.

5.7 Propagation effects

If the field point is located on the ground surface, the lightning channel is vertical, and the ground is assumed to be perfectly conducting, only two field components exist: the vertical electric field and the azimuthal magnetic field. The horizontal electric field component is zero, as required by the boundary condition on the surface of a perfect conductor. At an observation point above a perfectly conducting ground, a nonzero horizontal electric field component exists (see Appendix 3). A horizontal electric field exists both above ground and on (and below) its surface in the case of a finite ground conductivity. The horizontal (radial)

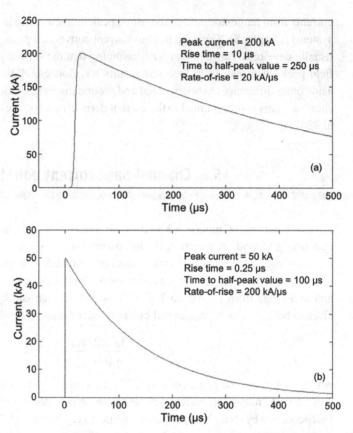

Fig. 5.3   Current waveforms recommended by the IEC 62305–1 (2010) for (a) first and (b) subsequent return strokes. The rise-time was measured between the 10% and 90% of peak value levels on the front part of the waveform. The rate-of-rise is the ratio of 0.8 of the peak value and 10–90% rise-time. The time to half-peak value was measured between the peak and half-peak value on the tail part of the waveform. Drawing by Vijaya B. Somu and Potao Sun.

electric field at and below the ground surface is associated with a radial current flow and resultant ohmic losses in the earth. A nonzero horizontal electric field on the ground surface makes the total electric field vector tilted relative to the vertical. The tilt is in the direction of propagation if the vertical electric field component is directed upward and in the direction opposite to the propagation direction if the vertical electric field component is directed downward, with the vertical component of the Poynting vector (power flow per unit area) being directed into the ground in both cases. Propagation effects include the preferential attenuation of the higher-frequency components in the vertical electric field and the azimuthal magnetic field waveforms.

Aoki et al. (2015) studied in detail the effects of finite ground conductivity on lightning electric and magnetic fields, using the 2D finite-difference time-domain method. Their distance range was from 5 to 200 km and the ground conductivity range from 0.1 mS/m (10^{-4} S/m) to infinity. They used the MTLL return-stroke model (Chapter 6) and considered the influence of source parameters, including the return-stroke speed v (ranging from c/2 to c) and current rise-time RT (ranging from 0.5 to 5 μs). The main results can be summarized as follows. The peaks of E_z, E_h (horizontal component of electric field), and B_φ are each nearly proportional to v. The peak of E_h decreases with increasing RT, while the peaks of E_z and B_φ are only slightly influenced by this parameter. At a distance of 5 km, the peaks of E_z and B_φ are essentially not affected by ground conductivity. Indeed, the difference between the 10^{-4} S/m and ∞ cases is only 1 percent for E_z and about 4 percent for B_φ. At 50 km, E_z and B_φ each reduces by about 5 percent for 10^{-2} S/m and by about 30 percent for 10^{-4} S/m relative to the perfectly conducting ground case. As expected, E_h decreases with increasing ground conductivity and vanishes at perfect ground.

5.8 Summary

The electric field intensity, $\overline{E}$, and magnetic flux density, $\overline{B}$, produced by lightning are usually found using the scalar, ϕ, and vector, $\overline{A}$, potentials. The latter two are related to the source quantities, the volume charge density, ρ, and current density, $\overline{J}$, respectively. The source quantities must satisfy the continuity equation. Alternatively, one can specify $\overline{J}$, find $\overline{A}$, and then use the Lorentz condition to find ϕ. In this latter case, there is no need to specify ρ in computing electric fields. The two approaches are equivalent; that is, they produce identical total fields, although the individual electric field components are different. For the transmission line model, the sum of electrostatic and induction components of electric field at close ranges and the radiation field component at far ranges each varies approximately as 1/r, where r is the horizontal distance between the lightning channel base and the ground-level observation point, and each is proportional to the channel-base current. The ranges of validity of these approximations depend on return-stroke current waveshape and speed. As the return-stroke speed v approaches c (speed of light), the approximate close-range equation (derived using only the *electrostatic and induction* field components) and the approximate far-range equation (derived using only the *radiation* field component) converge to the same (exact) equation for the *total* electric field, which is valid for *any distance* from the lightning channel.

If the observation point is located on the ground surface, the lightning channel is vertical, and the ground is assumed to be perfectly conducting, only two field components exist: the vertical electric field and the azimuthal magnetic field. On the finitely-conducting ground, there will be also a horizontal (radial) electric field component, so that the total electric field vector will be tilted from the vertical.

Questions and problems

5.1. Name the individual field components in Eqs. 5.7 and 5.8 and discuss their dependences on source parameters and distance.

5.2. Derive the far magnetic field approximation (similar to Eq. 5.15 for the far electric field) based on the transmission line model. Start with Eq. 5.8. List all the assumptions.

5.3. The radiation component of electric field generated by a dipole is zero along the axis of the dipole. Why?

5.4. What is the expected electric field at ground 30 m from the lightning channel, if the measured field at 15 m is 100 kV/m?

5.5. Compare the radiation field components generated by the entire lightning channel and by the electrically short current element.

5.6. Discuss the field propagation effects due to finite ground conductivity.

5.7. For the transmission line model with return-stroke speed equal to the speed of light, the total electric and total magnetic fields each have the same waveshape as that of the channel-base current and vary as the inverse distance. Why?

5.8. Why do the electrostatic and induction components of the electric field produced by an electrically short current element exhibit the reversal distance, while the radiation component does not?

Further reading

Appendix 3: Derivation of exact equations for computing lightning electric and magnetic fields

Cooray, V. ed. (2012). *Lightning Electromagnetics*, 950 pp., London: IET.

Rakov, V. A. and Uman, M. A. (2003). *Lightning: Physics and Effects*, 687 pp., New York: Cambridge.

Uman, M. A. (1987). *The Lightning Discharge*, 377 pp., San Diego, California: Academic Press.

6 Modeling of the lightning return stroke

6.1 Introduction and classification of models

Any lightning model is a mathematical construct designed to reproduce certain aspects of the physical processes involved in the lightning discharge. No modeling is complete until model predictions are compared with experimental data; that is, model testing, often called validation, is a necessary component of any modeling. An overview of various models of the lightning return stroke is given below.

Four classes of lightning return stroke models are usually defined. Most published models can be assigned to one, or sometimes two, of these four classes. The classes are primarily distinguished by the type of governing equations:

(1) The first class of models is the gas dynamic or "physical" models, which are primarily concerned with the radial evolution of a short segment of the lightning channel and its associated shock wave. These models typically involve the solution of three gas dynamic equations representing the conservation of mass, of momentum, and of energy, coupled to two equations of state. Principal model outputs include temperature, pressure, and mass density as a function of the radial coordinate and time.

(2) The second class of models is the electromagnetic models. These models involve a numerical solution of Maxwell's equations to find the current distribution along the channel from which the remote electric and magnetic fields can be computed.

(3) The third class of models is the distributed-circuit models that can be viewed as an approximation to the electromagnetic models and that represent the lightning discharge as a transient process on a vertical transmission line characterized by resistance (R), inductance (L), and capacitance (C), all per unit length. The governing equations are the telegrapher's equations. The distributed-circuit models (also called R-L-C transmission line models) are used to determine the channel current versus time and height and can therefore be used for the computation of remote electric and magnetic fields.

(4) The fourth class of models is the engineering models in which a spatial and temporal distribution of the channel current (or the channel line charge density) is specified based on such observed lightning return-stroke characteristics as current at the channel base, the speed of the upward-propagating front, and the channel luminosity profile. In these models, the physics of the lightning return stroke is deliberately downplayed, and the emphasis is placed on achieving agreement between the model-predicted

electromagnetic fields and those observed at distances from tens of meters to hundreds of kilometers. A characteristic feature of the engineering models is the small number of adjustable parameters, usually only one or two besides the specified channel-base current.

Outputs of the electromagnetic, distributed-circuit, and engineering models can be used directly for the computation of electromagnetic fields, while the gas dynamic models can be used for finding R as a function of time, which is one of the parameters of the electromagnetic and distributed-circuit models. Since the distributed-circuit and engineering models generally do not consider lightning channel branches, they best describe subsequent strokes or first strokes before the first major branch has been reached by the upward-propagating return stroke front, a time that is usually longer than the time required for the formation of the initial current peak at ground level.

Engineering models are discussed in more detail in the following sections of this chapter.

6.2 Engineering models

An engineering return-stroke model, as defined here, is simply an equation relating the longitudinal channel current $I(z',t)$ at any height z' and any time t to the current $I(0,t)$ at the channel origin, $z' = 0$. An equivalent expression in terms of the line charge density $\rho_L(z',t)$ on the channel can be obtained using the continuity equation (Thottappillil et al., 1997). Thottappillil et al. (1997) defined two components of the charge density at a given channel section, one component being associated with the return-stroke charge transferred through the channel section and the other with the charge deposited at the channel section. As a result, their charge density formulation reveals new aspects of the physical mechanisms behind the models (for example, the existence of radial current associated with the neutralization of leader charge stored in the corona sheath) that are not apparent in the longitudinal-current formulation. We first consider the mathematical and graphical representations of some simple models and then categorize and discuss the most used engineering models based on their implications regarding the principal mechanism of the return-stroke process. Rakov (1997) expressed several engineering models by the following generalized current equation:

$$I(z',t) = u(t - z'/v_f)P(z')I(0, t - z'/v) \tag{6.1}$$

where u is the Heaviside function equal to unity for $t \geq z'/v_f$ and zero otherwise, $P(z')$ is the height-dependent current attenuation factor, v_f is the upward-propagating return-stroke front speed, and v is the current-wave propagation speed. Table 6.1 summarizes $P(z')$ and v for five engineering models, namely, the transmission line model, TL (not to be confused with the R-L-C transmission line models); the modified transmission line model with linear current decay with height, MTLL; the modified transmission line model with exponential current decay with height, MTLE; the Bruce-Golde model, BG; and the traveling current source model, TCS. In Table 6.1, H is the total channel height, λ is the current decay height

Table 6.1 $P(z')$ and v in Eq. 6.1 for five engineering models		
Model	$P(z')$	v
TL	1	v_f
(Uman and McLain, 1969)		
MTLL	$1 - z'/H$	v_f
(Rakov and Dulzon, 1987)		
MTLE	$\exp(-z'/\lambda)$	v_f
(Nucci et al., 1988)		
BG	1	∞
(Bruce and Golde, 1941)		
TCS	1	$-c$
(Heidler, 1985)		

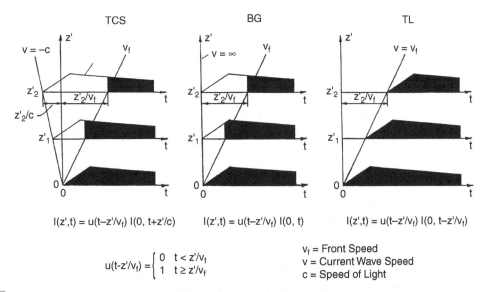

TCS

$I(z',t) = u(t-z'/v_f)\, I(0,\, t+z'/c)$

BG

$I(z',t) = u(t-z'/v_f)\, I(0,\, t)$

TL

$I(z',t) = u(t-z'/v_f)\, I(0,\, t-z'/v_f)$

$$u(t-z'/v_f) = \begin{cases} 0 & t < z'/v_f \\ 1 & t \geq z'/v_f \end{cases}$$

v_f = Front Speed
v = Current Wave Speed
c = Speed of Light

Fig. 6.1 Current versus time waveforms at ground ($z' = 0$) and at two heights z'_1 and z'_2 above ground for the TCS, BG, and TL return-stroke models. Slanted lines labeled v_f represent upward speed of the return-stroke front, and lines labeled v represent speed of the return-stroke current wave. The dark portion of the waveform indicates current that actually flows through a given channel section. Note that the current waveform at $z' = 0$ and v_f are the same for all three models. Adapted from Rakov (1997).

constant (assumed by Nucci et al. (1988a) to be 2000 m), and c is the speed of light. If not specified otherwise, v_f is assumed to be constant. The three simplest models, TCS, BG, and TL, are illustrated in Fig. 6.1 and the TCS and TL models additionally in Fig. 6.2. We consider first Fig. 6.1. For all three models we assume the same current waveform at the channel base ($z' = 0$) and the same front speed represented in $z' - t$ coordinates by the slanted

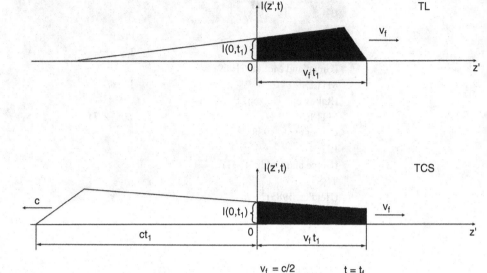

Fig. 6.2 Current versus height z′ above ground at an arbitrary fixed instant of time t = t₁ for the TL and TCS models. Note that the current at z′ = 0 and vf are the same for both models. Adapted from Rakov (1997).

line labeled v_f. The current-wave speed is represented by the line labeled v which coincides with the vertical axis for the BG model and with the v_f line for the TL model. Shown for each model are current versus time waveforms at the channel base ($z' = 0$) and at heights z'_1 and z'_2. Because of the finite front propagation speed v_f, current at a height, say, z'_2 begins with a delay z'_2/v_f with respect to the current at the channel base. The dark portion of the waveform indicates the current that actually flows through a given channel section, the blank portion being shown for illustrative purpose only. As seen in Fig. 6.1, the TCS, BG, and TL models are characterized by different current profiles along the channel, the difference being, from a mathematical point of view, due to the use of different values of v (listed in Table 6.1) in the generalized Eq. 6.1 with P(z′) = 1. It also follows from Fig. 6.1 that if the channel-base current were a step function, the TCS, BG, and TL models would be characterized by the same current profile along the channel, although established in an apparently different way in each of the three models. The relation between the TL and TCS models is further illustrated in Fig. 6.2 which shows that the spatial current wave moves in the positive z′ direction for the TL model and in the negative z′ direction for the TCS model. Note that in Fig. 6.2 the current at ground level (z′ = 0) and the upward moving front speed v_f are the same for both the TL and the TCS models. As in Fig. 6.1, the dark portion of the waveform indicates the current that actually flows in the channel, the blank portion being shown for illustrative purpose only.

The most used engineering models can be grouped in two categories: the transmission-line-type models and the traveling-current-source-type models, summarized in Tables 6.2 and 6.3, respectively. Each model in Tables 6.2 and 6.3 is represented by both current and

Table 6.2 Transmission-line-type models for $t \geq z'/v_f$.
$$Q(z',t) = \int_{z'/v}^{t} I(0, \tau - z'/v) d\tau, \quad v = v_f = \text{const}, \quad H = \text{const}, \quad \lambda = \text{const}$$

TL (Uman and McLain, 1969)	$I(z',t) = I(0, t - z'/v)$ $\rho_L(z',t) = \dfrac{I(0, t - z'/v)}{v}$
MTLL (Rakov and Dulzon, 1987)	$I(z',t) = \left(1 - \dfrac{z'}{H}\right) I(0, t - z'/v)$ $\rho_L(z',t) = \left(1 - \dfrac{z'}{H}\right)\left(\dfrac{I(0, t - z'/v)}{v}\right) + \dfrac{Q(z',t)}{H}$
MTLE (Nucci et al., 1988)	$I(z',t) = e^{-z'/\lambda} I(0, t - z'/v)$ $\rho_L(z',t) = e^{-z'/\lambda}\dfrac{I(0, t - z'/v)}{v} + \dfrac{e^{-z'/\lambda}}{\lambda} Q(z',t)$

Table 6.3 Traveling-current-source-type models for $t \geq z'/v_f$.
$$v^* = v_f/(1 + v_f/c), \quad v_f = \text{const}, \quad \tau_D = \text{const}$$

BG (Bruce and Golde, 1941)	$I(z',t) = I(0,t)$ $\rho_L(z',t) = \dfrac{I(0, z'/v_f)}{v_f}$
TCS (Heidler, 1985)	$I(z',t) = I(0, t + z'/c)$ $\rho_L(z',t) = -\dfrac{I(0, t + z'/c)}{c} + \dfrac{I(0, z'/v^*)}{v^*}$
DU (Diendorfer and Uman, 1990)	$I(z',t) = I(0, \, t + z'/c) - e^{-(t - z'/v_f)\tau_D^{-1}} I(0, \, z'/v^*)$ $\rho_L(z',t) = -\dfrac{I(0, \, t + z'/c)}{c} - e^{-(t - z'/v_f)\tau_D^{-1}}\left[\dfrac{I(0, \, z'/v^*)}{v_f} + \dfrac{\tau_D}{v^*}\dfrac{dI(0, \, z'/v^*)}{dt}\right]$ $\qquad + \dfrac{I(0, \, z'/v^*)}{v^*} + \dfrac{\tau_D}{v^*}\dfrac{dI(0, \, z'/v^*)}{dt}$

line charge density equations. Table 6.2 includes the TL model and its two modifications: the MTLL and MTLE models. The transmission-line-type models can be viewed as incorporating a current source at the channel base which injects a specified current wave into the channel, that wave propagating upward (1) without either distortion or attenuation (TL), or (2) without distortion but with specified attenuation (MTLL and MTLE), as seen from the corresponding current equations given in Table 6.2. The TL model is often portrayed as being equivalent to an ideal (lossless), uniform transmission line, which is not accurate. First, for such a transmission line in air the wave propagation speed is equal to the speed of light, while in the TL model it is set to a lower value, in order to make it consistent with observations. Secondly, a vertical conductor of nonzero radius above ground is actually a nonuniform transmission line whose characteristic impedance increases with increasing height. The resultant distributed impedance discontinuity causes distributed reflections back to the source that is located at ground level. For this reason, even in the absence of ohmic losses, the current amplitude appears to decrease with increasing height (Baba and Rakov, 2005b), the effect neglected in the TL model.

Table 6.3 includes the BG model, the TCS model, and the Diendorfer–Uman (DU) model. In the traveling-current-source-type models, the return-stroke current may be viewed as generated at the upward-moving return-stroke front and propagating downward or, equivalently, as resulting from the cumulative effect of shunt current sources that are distributed along the lightning channel and progressively activated by the upward-moving return-stroke front. In the TCS model, current at a given channel section turns on instantaneously as the front passes this section, while in the DU model, the current turns on gradually (exponentially with a time constant τ_D if $I(0,t+z'/c)$ were a step function). The channel current in the TCS model may be viewed as a single downward-propagating wave, as illustrated in Fig. 6.2. The DU model involves two terms (see Table 6.3), one being the same as the downward-propagating current in the TCS model that exhibits an inherent discontinuity at the upward-moving front (see Fig. 6.1), and the other being an opposite polarity current which rises instantaneously to a value equal in magnitude to the current at the front and then decays exponentially with a time constant τ_D. The second current component in the DU model may be viewed as a "front modifier." It propagates upward with the front and eliminates any current discontinuity at that front. The time constant τ_D is the time during which the charge per unit length deposited at a given channel section by the preceding leader reduces to $1/e$ (about 37 percent) of its original value after this channel section is passed by the upward-moving front. Thottappillil et al. (1997) assumed that $\tau_D = 0.1$ μs. Diendorfer and Uman (1990) considered two components of charge density, each released with its own time constant in order to match model predicted fields with measured fields. If $\tau_D = 0$, the DU model reduces to the TCS model. In both the TCS and DU models, the downward-propagating current wave speed is set to be equal to the speed of light. The TCS model reduces to the BG model if the downward current propagation speed is set equal to infinity instead of the speed of light. Although the BG model could be also viewed mathematically as a special case of the TL model with v replaced by infinity, we choose to include the BG model in the traveling-current-source-type model category.

The principal distinction between the two types of engineering models formulated in terms of current is the direction of the propagation of the current wave: upward for the transmission-line-type models ($v = v_f$) and downward for the traveling-current-source-type models ($v = -c$), as seen for the TL and TCS models, respectively, in Fig. 6.2. As noted above, the BG model can be viewed mathematically as a special case of either the TCS or the TL model. The BG model includes a current wave propagating at an infinitely large speed and, as a result, the wave's direction of propagation is indeterminate. As in all other models, the BG model includes a front moving at a finite speed v_f. Note that, even though the direction of propagation of the current wave in a model can be either up or down, the direction of current is the same; that is, charge of the same sign is transported to ground in both types of engineering models.

The TL model predicts that, as long as (1) the height above ground of the upward-moving return-stroke front is much smaller than the distance r between the observation point on the ground and the channel base, so that all contributing channel points are essentially equidistant from the observer, (2) the return-stroke front propagates at a constant speed, (3) the return-stroke front has not reached the top of the channel, and (4) the ground conductivity is high enough that the associated propagation effects are negligible, the vertical component

E_z^{rad} of the electric radiation field and the azimuthal component of the magnetic radiation field are each proportional to the channel-base current I (e.g. Uman et al., 1975). The equation for the electric radiation field E_z^{rad} is as follows:

$$E_z^{rad}(t) = -\frac{v}{2\pi\varepsilon_0 c^2 r}I(0, \, t - r/c) \tag{6.2}$$

where ε_0 is the permittivity of free space, v is the upward propagation speed of the current wave, which is the same as the front speed v_f in the TL model, and c is the speed of light. Eq. 6.2 is derived in Chapter 5 and is the same as Eq. 5.15. It is also found in Table 5.2 as the far-range approximation for E_z. For the most common return stroke lowering negative charge to ground, the sense of the positive charge flow is upward so that current I, assumed to be upward-directed in deriving Eq. 6.2, by convention is positive, and E_z^{rad} by Eq. 6.2 is negative; that is, the electric field vector points in the negative z direction. Taking the derivative of this equation with respect to time, one obtains

$$\frac{\partial E_z^{rad}(r, t)}{\partial t} = -\frac{v}{2\pi\varepsilon_0 c^2 r}\frac{\partial I(0, \, t - r/c)}{\partial t} \tag{6.3}$$

Equations 6.2 and 6.3 are commonly used, particularly the first one and its magnetic radiation field counterpart, found from $|B_\phi^{rad}| = |E_z^{rad}|/c$, where B_ϕ^{rad} is the radiation component of the azimuthal magnetic flux density, for the estimation of the peak values of return-stroke current and its time derivative, subject to the assumptions listed prior to Eq. 6.2. General equations for computing electric and magnetic fields at ground are derived in Appendix 3 and the final results are also found in Chapter 5.

6.3 Equivalency between the lumped-source and distributed-source representations

Maslowski and Rakov (2007) showed that any engineering return-stroke model can be expressed, using an appropriate continuity equation, in terms of either a lumped current source placed at the bottom of the channel or multiple shunt current sources distributed along the channel, with the resultant longitudinal current and the total charge density distribution along the channel being the same in both formulations. This property can be viewed as the duality of engineering models.

In general, any engineering model includes (explicitly or implicitly) both the longitudinal and transverse (radial) currents, with the radial current in the TL model being zero. This is illustrated in Fig. 6.3. In the distributed-source-type (DS) models, the radial current, supplied by distributed current sources, enters the channel, with the current sink being at the bottom of the channel. In the lumped-source-type (LS) models, the radial current leaves the channel to compensate the leader charge stored in the corona sheath; the current source is at the bottom of the channel and the partial radial currents can be viewed as current sinks distributed along the channel. Although the directions of the actual radial current in the LS and DS models are opposite (out of the channel and into the channel, respectively), charge of the same sign is

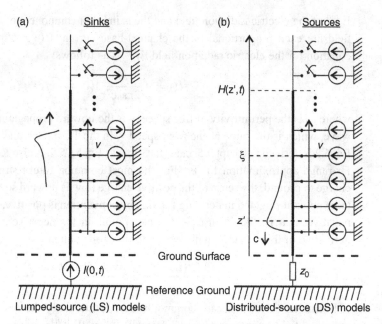

Fig. 6.3   Schematic representation of engineering return-stroke models that employ (a) a lumped current source at the lightning channel base (LS models) and (b) distributed current sources along the channel (DS models). v here is the upward return-stroke front speed, c is the speed of light, and Z_0 is the characteristic impedance of the lightning channel (matched conditions at ground are implied in DS models). LS models with longitudinal-current decay with height imply current sinks distributed along the channel, as shown in (a). Adapted from Maslowski and Rakov (2007).

effectively transported into the channel core in both types of models. It is also important to note that for either type of model the radial current is distributed along the channel.

Conversion of, for example, an LS model to equivalent DS model amounts to replacement of the actual corona current of the model by an equivalent one that results in a reversal of direction of the longitudinal current (the source becomes a sink and the sinks become sources), while the longitudinal current and the total charge density distribution along the channel remain the same. For the TL model (no longitudinal-current attenuation with height and, hence, no radial current) conversion from the LS to equivalent DS formulation leads to the introduction of a purely fictitious bipolar radial current which is required to make the lumped source a sink (see Fig. 6.3a). Conversion of a DS model to equivalent LS model is done in a similar manner. Conversion of LS models to DS models is particularly useful in extending return-stroke models to include a tall strike object (see Section 6.4), as done, for example, by Rachidi et al. (2002).

6.4 Extension of models to include a tall strike object

Some engineering models have been extended to include a grounded strike object modeled as an ideal, uniform transmission line that supports the propagation of waves at the speed of light without attenuation or distortion. Such an extension results in a second current wavefront which propagates from the top of the object toward ground at the speed of light, produces reflection on its arrival there, and is allowed to bounce between the top and bottom ends of the object and, in general, produce transmitted waves at either end. The transient behavior of tall objects under direct lightning strike conditions can be illustrated as follows. For the simple example of a non-ideal current source (ideal current source in parallel with source impedance) attached to the top of the object (see Fig. 4.30b) generating a step-function current wave, the magnitude of the wave injected into the object depends on the characteristic impedance of the object. Specifically, the ideal source current initially divides between the source impedance and the characteristic impedance of the object. However, after a sufficiently long period of time, the current magnitude at any point on the object will be equal to the magnitude of current that would be injected directly into the grounding impedance of the object from the same current source in the absence of the object. In other words, at late times, the ideal source current will in effect divide between the source impedance and the grounding impedance, as if the strike object were not there. Note that the above example applies only to a step-function current wave, the current distribution along the object being more complex for the case of an impulsive current waveform characteristic of the lightning return stroke. If the lightning current wave round-trip time on the strike object is appreciably longer than the rise time of current measured at the top of the object, the current peak reflected from the ground is separated from the incident-current peak in the overall current waveform, at least in the upper part of the object.

Model-predicted lightning electromagnetic environment in the presence of tall (electrically long) strike object was studied by many researchers. According to Baba and Rakov (2007b), for a typical subsequent stroke, the vertical electric field due to a lightning strike to a 100 m-high object is expected to be reduced relative to that due to the same strike to flat ground at distances ranging from 30 m to 200 m from the object and enhanced at distances greater than 200 m. The azimuthal magnetic field for the tall object case is larger than that for the flat ground case at any distance. Beyond about 3 km the field peak is essentially determined by its radiation component and the so-called far-field enhancement factor becomes insensitive to distance change and expected to be equal to about 2.3.

Note that when the shortest significant wavelength in the lightning current is much longer than the height of the strike object, there is no need to consider the distributed-circuit behavior of such an object. For example, if the minimum significant wavelength is 300 m (corresponding to a frequency of 1 MHz), objects whose heights are about 30 m or less may be considered as lumped, in most cases as a short circuit between the lightning channel base and grounding impedance of the object (see Fig. 4.30a).

6.5 Testing model validity

The overall strategy in testing the validity of engineering models is illustrated in Fig. 6.4. For a given set of input parameters, including $I(0,t)$ and v_f, the model is used to find the distribution of current $I(z',t)$ along the channel, which is then used for computing electric and magnetic fields (using equations found in Chapter 5) at different distances from the

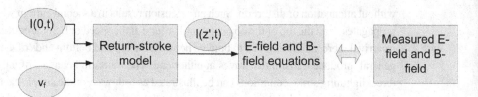

Fig. 6.4 Illustration of the overall strategy in testing the validity of engineering models. Artwork by Potao Sun.

lightning channel. The model-predicted fields are compared with corresponding measured fields. Current and field derivatives can be also used for model testing. Ideally, a good model should reproduce measurements at all distances for which the employed electric and magnetic field equations are valid. However, in practice it may be sufficient for a model to be able to reasonably reproduce measured fields only within a certain range of distances (for example, only at far distances when the fields are dominated by their radiation components) or/and for certain times (for example, only for the first few microseconds, when the current and field peaks usually occur).

Two primary approaches to model testing have been used: the so-called "typical-event" and "individual-event" approaches. The "typical-event" approach involves the use of a *typical* channel-base current waveform $I(0,t)$ and a *typical* front propagation speed v_f as inputs to the model and a comparison of the model-predicted fields with *typical* observed fields. In the "individual-event" approach, $I(0,t)$ and v_f, both measured for the same individual event are used to compute fields that are compared to the measured fields for that same event. When v_f is not available, a range of reasonable values (measured values are typically in the $c/3$ to $2c/3$ range; see Chapter 4) can be used to see if a match with measured fields can be achieved for any of the speed values. The individual-event approach is able to provide a more definitive answer regarding model validity, but it is feasible only in the case of triggered-lightning return strokes or natural lightning strikes to tall towers where channel-base current can be measured.

In the field calculations, the channel is generally assumed to be straight and vertical with its origin at ground level ($z' = 0$): conditions which are expected to be valid for subsequent strokes, but not necessarily for first strokes. The channel length is usually not specified unless it is an inherent feature of the model, as is the case for the MTLL model. As a result, the model-predicted fields and associated model validation may not be meaningful after 25–75 μs, the expected time it takes for the return-stroke front to traverse the distance from ground to the cloud charge source region.

6.5.1 "Typical-event" approach

This approach has been adopted by Nucci et al. (1990), Rakov and Dulzon (1991), and Thottappillil et al. (1997) among others.

Distant (1 to 200 km) fields. Nucci et al. (1990) identified four characteristic features in the fields at 1 to 200 km measured by Lin et al. (1979) (Fig. 4.32) and used those features as

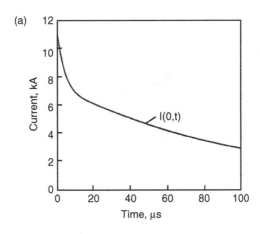

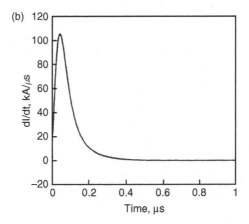

Fig. 6.5 (a) Current at ground level and (b) corresponding current derivative used by Nucci et al. (1990), Rakov and Dulzon (1991), and Thottappillil et al. (1997) for testing the validity of return-stroke models by means of the "typical-event" approach. The peak current is about 11 kA, and peak current rate of rise is about 105 kA/μs. Adapted from Nucci et al. (1990).

a benchmark for their validation of the TL, MTLE, BG, and TCS models (also of the MULS model, not considered here). The characteristic features include (1) a sharp initial peak that varies approximately as the inverse distance beyond a kilometer or so in both electric and magnetic fields, (2) a slow ramp following the initial peak and lasting in excess of 100 μs for electric fields measured within a few tens of kilometers, (3) a hump following the initial peak in magnetic fields within a few tens of kilometers, the maximum of which occurs between 10 and 40 μs, and (4) a zero crossing within tens of microseconds of the initial peak in both electric and magnetic fields at 50 to 200 km. For the current (Fig. 6.5) and other model characteristics assumed by Nucci et al. (1990), feature (1) is reproduced by all the models examined, feature (2) by all the models except for the TL model, feature (3) by the BG, TL, and TCS models, but not by the MTLE model, and feature (4) only by the MTLE

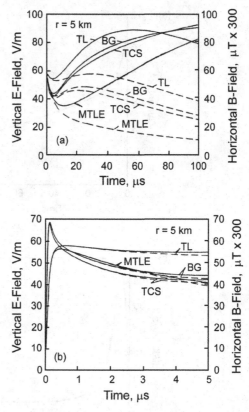

Fig. 6.6 Calculated vertical electric (left scaling, solid lines) and horizontal (azimuthal) magnetic (right scaling, dashed lines) fields for four return-stroke models at a distance r = 5 km displayed on (a) 100 μs and (b) 5 μs time scales. Adapted from Nucci et al. (1990).

model, but not by the BG, TL, and TCS models, as illustrated in Figs. 6.6 and 6.7. Diendorfer and Uman (1990) showed that the DU model reproduces features (1), (2), and (3), and Thottappillil et al. (1991b) demonstrated that a relatively insignificant change in the channel-base current waveform (well within the range of typical waveforms) allows the reproduction of feature (4), the zero crossing, by the TCS and DU models. Rakov and Dulzon (1991) showed that the MTLL model reproduces features (1), (2), and (4). Nucci et al. (1990) conclude from their study that all the models evaluated by them using measured fields at distances ranging from 1 to 200 km predict reasonable fields for the first 5–10 μs, and all models, except the TL model, do so for the first 100 μs.

There is another "typical-event" method of testing the validity of return-stroke models, which is based on using net electrostatic field changes produced by leader, ΔE_L, and return-stroke, ΔE_{RS}, processes. The ratio of these field changes, $\Delta E_L / \Delta E_{RS}$, depends on the distribution of charge along the fully formed leader channel. For a uniformly charged channel, this ratio at far distances is equal to 1 (see Fig. 4.16), if one assumes that the return stroke completely neutralizes the leader charge and deposits no additional charge

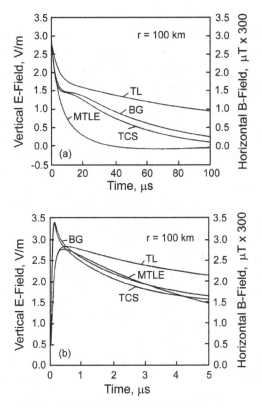

Fig. 6.7 Same as Fig. 6.6, but for r = 100 km. Adapted from Nucci et al. (1990).

anywhere in the system (this assumption is discussed later in this section). If the leader charge density distribution is skewed toward the ground, the ratio will be greater than 1, and smaller than 1 if it is skewed toward the cloud. Now, the leader charge density distribution is related to the return-stroke current decay along the channel (Rakov and Dulzon, 1991). Specifically, the uniform charge density distribution corresponds to a linear current decay with height (MTLL model), a linear charge density decrease with increasing height to a parabolic current decay with height, and an exponential charge density decrease with increasing height to an exponential current decay with height (MTLE model). In other words, a return-stroke model predicting a variation of current magnitude with height also implicitly specifies the distribution of leader charge density along the channel. It follows that computing $\Delta E_L / \Delta E_{RS}$ at a far distance (some tens of kilometers) for different return-stroke models and comparing model predictions with measurements can be used for testing model validity. Thottappillil et al. (1997, Table 2) assumed that the charge source height is 7.5 km and found that at 100 km, the field change ratio is equal to 0.99 (0.81 at 20 km) for a uniformly charged leader (MTLL model) and 3.1 (2.6 at 20 km) for a leader with charge density exponentially decaying with height (MTLE model). Close to 1 ratios were also computed by Thottappillil et al. (1997) for the BG, TCS, and DU models. These results are to be compared to the observations of Beasley et al. (1982, Fig. 23d), who, for 97 first strokes at distances of approximately 20 to 50 km, reported a mean value for the ratio of 0.8 (see also Fig. 4.17). It follows that the BG, MTLL, TCS, and DU models are supported by the available $\Delta E_L/\Delta E_{RS}$ measurements at far ranges from the lightning channel, whereas the MTLE model (whose leader charge density distribution is strongly skewed toward ground) is not.

The $\Delta E_L/\Delta E_{RS}$ test can be also applied to the bidirectional leader model of Mazur and Ruhnke (1993), who simulated the upper section of the leader system by a single-channel positive leader extending vertically upward at the same speed as the negative section of the leader system. They found that the charge density on such a vertically symmetrical bidirectional leader, regarded as a vertical conductor polarized in a uniform electric field, varied linearly with height, with zero charge density at the origin. The model of Mazur and Ruhnke predicts $\Delta E_L/\Delta E_{RS}$ at 20 to 50 km to be between approximately 0.2 and 0.3 (see their Fig. 25), which is significantly lower than the average value of 0.8 observed for first strokes at these distances by Beasley et al. (1982).

In calculations of the $\Delta E_L/\Delta E_{RS}$ ratio discussed above, the leader charge is assumed to be exactly equal to the return-stroke charge. In general, the total positive charge that enters the leader channel at the strike point (or the negative charge that goes into the ground) during the return stroke can be divided into three components. The first part is the positive charge that is necessary to neutralize the negative charge originally stored in the leader channel. The second part is the positive charge induced in the return stroke channel, which is essentially at ground potential, due to the background electric field produced by remaining negative cloud charges. The third part is the additional positive charge spent to neutralize negative cloud charge that was not involved in the leader process. Note that the return stroke charge is often defined as the first part only (as done in calculations of the $\Delta E_L/\Delta E_{RS}$ ratio discussed above), while all three parts can materially contribute to the measured electric field change. Thus, some caution is to be exercised in using the $\Delta E_L/\Delta E_{RS}$ ratio method of testing model validity. Note also that this method is not applicable to the TL model, in which the implicit leader charge is zero.

Close (tens to hundreds of meters) fields. Thottappillil et al. (1997) noted that measured electric fields at tens to hundreds of meters from triggered lightning exhibit a characteristic flattening within 15 μs or so, as seen in Fig. 4.37. Electric fields predicted at 50 m by the BG, TL, MTLL, TCS, MTLE, and DU models are shown in Fig. 6.8. As follows from this figure, the BG, MTLL, TCS, and DU models, but not the TL and MTLE models, more or less reproduce the characteristic field flattening at later times seen in Fig. 4.37. Thus, the TL and MTLE models should be viewed as inadequate for computing close electric fields at later times (beyond the initial 10 μs or so).

6.5.2 "Individual-event" approach

This approach has been adopted by Thottappillil and Uman (1993) who compared the TL, TCS, MTLE, and DU models (also the MDU model, not considered here) and Schoene et al. (2003) who compared the TL and TCS models. Thottappillil and Uman (1993) used 18 sets of three simultaneously-measured features of triggered-lightning return strokes: channel-base

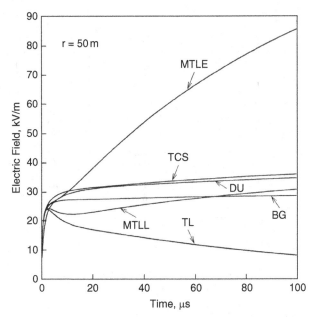

Fig. 6.8 Calculated vertical electric fields for six return-stroke models at a distance r = 50 m, to be compared with typical measured return-stroke field at 50 m presented in Fig. 4.37. Note that only the upward-going portion of the waveforms shown in Fig. 4.37 is due to the return stroke, the downward-going portion being due to the preceding dart leader. Adapted from Thottappillil et al. (1997).

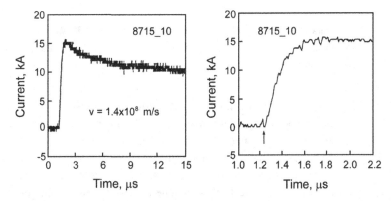

Fig. 6.9 An example of current waveform at the base of the channel (left-hand panel) and a close-up of the current wave front on an expanded timescale (right-hand panel) used by Thottappillil and Uman (1993) for testing the validity of return-stroke models by means of the "individual-event" approach. Also given is the measured return-stroke speed. Adapted from Thottappillil and Uman (1993).

current, return-stroke propagation speed, and electric field at about 5 km from the channel base, all obtained by Willett et al. (1989a). An example of the comparison for one return stroke, whose measured channel-base current and measured speed are found in Fig. 6.9, is given in Fig. 6.10. It was found for the overall data set that the TL, MTLE, and DU models

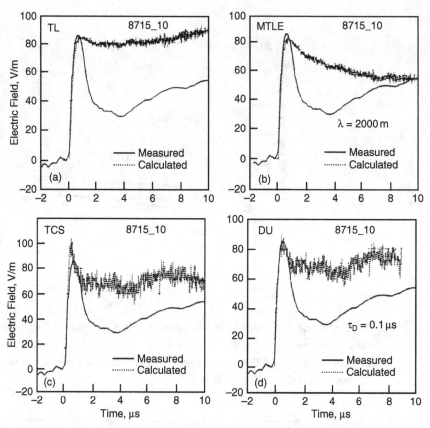

The vertical electric fields ("noisy" dotted lines) calculated using the TL, MTLE, TCS, and DU models, shown together with the measured field (solid lines) at about 5 km for the return stroke whose measured current at the channel base and the measured return-stroke speed are given in Fig. 6.9. Adapted from Thottappillil and Uman (1993).

each predicted the measured initial electric field peaks with an error whose mean absolute value was about 20 percent, while the TCS model had a mean absolute error of about 40 percent. From the standpoint of the overall field waveforms at 5 km, all the tested models should be considered less than adequate.

Schoene et al. (2003b) tested the TL and TCS models by comparing the first microsecond of model-predicted electric and magnetic field waveforms and field derivative waveforms at 15 m and 30 m with the corresponding measured waveforms from triggered-lightning return strokes. The electric and magnetic fields were calculated from Eqs. 5.7 and 5.8 given the measured current or current derivative at the channel base, an assumed return stroke front speed (three values, $v = 1 \times 10^8$ m/s, $v = 2 \times 10^8$ m/s, and $v = 2.99 \times 10^8$ m/s, were considered), and the temporal and spatial distribution of the channel current specified by the return-stroke model. This was a somewhat lesser testing method than the "individual approach" discussed above, since the speeds were not measured and had to be assumed. The TL model was found to work reasonably well in predicting the measured electric and magnetic fields (and field derivatives) at 15 m and 30 m if return stroke speeds during the first microsecond were chosen to be between 1×10^8 m/s and 2×10^8 m/s. The TCS model did not adequately predict either the measured electric fields or the measured electric and magnetic field derivatives at 15 and 30 m during the first microsecond or so. The TCS model deficiency is related to the fact that it implicitly assumes that the channel is terminated in its characteristic impedance (see Fig. 6.3b); that is, the current reflection coefficient at ground level is zero. In most cases this assumption is invalid since the impedance of the lightning channel is typically much larger than the impedance of the grounding, resulting in a current reflection coefficient close to one (there are close to short circuit, rather than matched conditions at ground).

Based on the entirety of the testing results and mathematical simplicity, the engineering models are ranked in the following descending order: MTLL, DU, MTLE, TCS, BG, and TL. However, the TL model is recommended for the estimation of the initial field peak from the current peak or conversely the current peak from the field peak, since it is the mathematically simplest model with a predicted peak field/peak current relation that is not less accurate than those of the more mathematically complex models.

6.6 Summary

Lightning return-stroke models can be assigned to one, sometimes two, of the following four classes: (1) gas-dynamic models, (2) electromagnetic models, (3) distributed-circuit models, and (4) engineering models. The most used engineering models can be grouped in two categories: the transmission-line-type models (lumped current source at the bottom of the channel) and the traveling-current-source-type models (multiple equivalent current sources distributed along the channel). Any lumped-source model can be converted to its equivalent distributed-source model and vice versa, with the resultant longitudinal current and total charge density distribution along the channel being the same. This property can be viewed as the duality of engineering models. Conversion of lumped-source models to distributed-source models is particularly useful in extending return-stroke models to include a tall strike object. Testing model validity is a necessary component of modeling. The engineering models are most conveniently tested using measured electric and magnetic fields from natural and triggered lightning. Based on the entirety of the testing results and on mathematical simplicity, the engineering models are ranked in the following descending order: MTLL, DU, MTLE, TCS, BG, and TL. When only the relation between the initial peak values of the channel-base current and the remote electric or magnetic fields is concerned, the TL model is preferred.

Questions and problems

6.1. Name the four classes of return-stroke models and their governing equations.
6.2. What is the difference between the lumped-source and distributed-source models?

6.3. How to test the validity of engineering return-stroke models?

6.4. The transmission line model cannot adequately reproduce the overall electric fields at either close or far distance (see Figs. 6.6–6.8). Nevertheless, it is often acceptable in practice. Why?

6.5. Why are the far lightning fields enhanced in the presence of tall strike object?

6.6. Which type of ideal source would you use in the return-stroke model extended to include a tall strike object: (a) series lumped current source, (b) shunt lumped current source, (c) series lumped voltage source, (d) shunt current sources distributed along the lightning channel? Explain why.

6.7. Why does the return-stroke current amplitude decrease with increasing height? Show that the modified transmission line model with linear current decay with height (MTLL) corresponds to a uniformly charged leader channel. Hint: P(z′) is the ratio of the integral of the line charge density from z′ to H and that from 0 to H.

6.8. Using the transmission line model, estimate the lightning peak current from the return-stroke electric field waveform shown in Fig. 4.33c. Note that all field waveforms in that figure are normalized to 100 km and that the return-stroke speed can be taken to be equal to 1.5×10^8 m/s (typical value).

Further reading

Baba, Y. and Rakov, V. A. (2015). *Electromagnetic Computation Methods for Lightning Surge Protection Studies*, 320 pp., Wiley-IEEE Press.

Cooray, V., ed. (2012). *Lightning Electromagnetics*, 950 pp., London: IET.

Cooray, V., ed. (2014). *The Lightning Flash*, 2nd Edition, 896 pp., London: IET.

Rakov, V. A. and Uman, M. A. (2003). *Lightning: Physics and Effects*, 687 pp., New York: Cambridge.

Uman, M. A. (1987). *The Lightning Discharge*, 377 pp., San Diego, California: Academic Press.

Uman, M. A. (2001). *The Lightning Discharge*, 377 pp., Mineola, New York: Dover.

Measurement of lightning electric and magnetic fields

Principles of measurement of electric and magnetic fields produced by the various lightning processes are introduced in this chapter. Equivalent circuits of the antennas and associated electronics (including the input impedance of the recorder) are discussed. Conditions for faithful reproduction by the measuring system of the lightning-generated field waveforms are considered. The focus here is on field measurements at ground level, where the electric field is dominated by its vertical component. Characteristics of wideband electric and magnetic fields produced by lightning, that are needed in designing field measuring systems, are given at the end of Section 7.1.3 (for electric field measurement in the immediate vicinity of the lightning channel) and in Section 7.3.

7.1 Wideband electric field measurements

7.1.1 Ordinary capacitive antennas

A sensor that is commonly used to measure the lightning vertical electric field is a metallic disk placed flush with the ground surface: the so-called flat-plate antenna. Figure 7.1a schematically shows such an antenna, where it is assumed that the area A of the antenna sensing plate is small enough to consider the electric field E constant over that area and C_a is the capacitance of the antenna. The downward-directed electric field induces negative charge Q on the surface of the antenna, which can be found as the product of the surface charge density ρ_s and the area A of the antenna sensing plate. From the boundary condition on the vertical component of electric field on the surface of good conductor, $\rho_s = \varepsilon_0 E$, where ε_0 is the electric permittivity of free space, and hence $Q = \varepsilon_0 EA$. If E is varying with time, there will be current $I = dQ/dt = \varepsilon_0 A dE/dt$ flowing between the sensing plate and ground. This current is proportional to dE/dt. In order to measure E, it is necessary to use an integrating capacitor $C \gg C_a$, (see Fig. 7.1b), since C_a is usually too small for measuring lightning fields, as will be discussed later in this section. Thus, the voltage across the integrating capacitor (capacitive voltage drop) will be

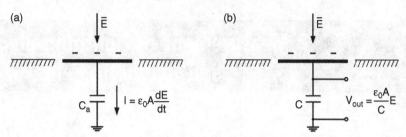

Fig. 7.1 Illustration of the principle of operation of the flat-plate antenna. (a) Antenna without external circuit. (b) Antenna with external integrating capacitor $C \gg C_a$. Drawing by Potao Sun.

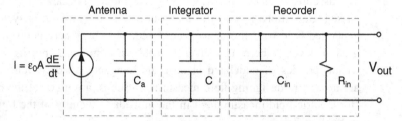

Fig. 7.2 Norton equivalent circuit of electric field antenna shown along with the integrating capacitance and the input impedance of recorder (usually $C \gg C_a$ and $C \gg C_{in}$). Drawing by Potao Sun.

$$V_{out} = \frac{1}{C_a + C} \int_0^t I(t')dt' \approx \frac{1}{C} \int_0^t I(t')dt' = \frac{Q}{C} = \frac{\varepsilon_0 AE}{C} \qquad (7.1)$$

Strictly speaking, Eq. 7.1 applies only to the case of infinitely large input impedance of the recorder. In practice, the input resistance of the recorder (or fiber-optic-link transmitter) plays an important role, limiting the time interval or the lower end of the frequency range over which Eq. 7.1 is valid. To examine this further, it is convenient to use the Norton equivalent circuit of the antenna, which is the antenna short-circuit current, $I = \varepsilon_0 Aj\omega E$ (ideal current source), in parallel with antenna impedance, $1/j\omega C_a$, where $\omega = 2\pi f$ with f being frequency in hertz. The equivalent circuit including the Norton equivalent of the antenna, integrating capacitance, and input resistance R_{in} and capacitance C_{in} of the recorder is shown in Fig. 7.2.

Since $C \gg C_a$ and usually $C \gg C_{in}$, the current basically splits between C and R_{in}, and Eq. 7.1 holds when $1/\omega C \ll R_{in}$; that is, $\omega \gg 1/(R_{in}C)$ or $f \gg 1/(2\pi R_{in}C)$. In the time domain, Eq. 7.1 is valid when the variation time (duration) of the signal of interest $\Delta t \ll \tau$, where $\tau = R_{in}C$ is the decay time constant of the measuring system (when E is a step-function, V_{out} will exponentially decay to $1/e$, where e is the base of the natural logarithm, or about 37 percent of its initial value over the time equal to τ). For example, if $C = 1$ μF and $R_{in} = 1$ MΩ, $\tau = 1$ s, long enough for recording electric fields produced by lightning

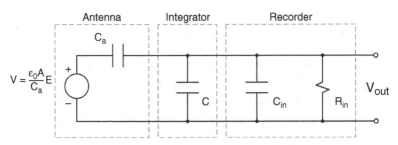

Fig. 7.3 Thevenin equivalent circuit of electric field antenna shown along with the integrating capacitance and the input impedance of recorder (usually $C \gg C_a$ and $C \gg C_{in}$). Drawing by Potao Sun.

processes occurring on timescales of the order of tens of milliseconds (for example, stepped leaders or return strokes followed by continuing currents). Typical values of C_a and C_{in} are of the order of tens to hundreds of picofarad (1 pF $= 10^{-12}$ F) or less, clearly much smaller than $C = 1$ μF (10^{-6} F) in this example. For recording return-stroke pulses, τ of the order of milliseconds is usually sufficient, while for the faithful reproduction of overall flash waveforms it should be of the order of 10 s or so. For recording microsecond- and submicrosecond-scale pulses, τ shorter than a millisecond or so can be used. Measuring systems with decay time constants of the order of seconds are sometimes referred to as "slow antenna" systems, and those with submillisecond time constants as "fast antenna" systems. Fast antenna systems usually have higher gains than slow antenna ones. The terms "slow" and "fast" have nothing to do with the upper frequency response of the system, which is usually determined by the amplifier or fiber-optic link. The measuring system shown in Fig. 7.2 employs a passive integrator. In the case of active integrator, $\tau = RC$ is determined by R and C connected in parallel in the feedback circuit of the operational amplifier. Sometimes, a Y-connected resistor network is used in the feedback circuit of an operational amplifier, which allows one to use resistors smaller than the required R value.

The antenna can also be represented by the Thevenin equivalent circuit, which is the antenna open-circuit voltage (ideal voltage source) in series with the antenna impedance (the latter being the same as in the Norton equivalent circuit). The open-circuit voltage of the antenna is the product of its short-circuit current and its impedance, $V = \varepsilon_0 Aj\omega E/(j\omega C_a) = \varepsilon_0 AE/C_a$. The corresponding overall equivalent circuit of the field measuring system is shown in Fig. 7.3. When $1/\omega C \ll R_{in}$, the ideal-source voltage is applied to the capacitive voltage divider formed by C_a and C (neglecting C_{in}), and $V_{out} = \varepsilon_0 AE/C$, as expected (the Norton and Thevenin representations of the antenna are equivalent).

We now discuss the situation when the condition of $\Delta t \ll \tau$ is not satisfied. Such situations are not rare. Indeed, since the range of lightning electric field changes is very large (it spans orders of magnitude), it is practically impossible to build a single measuring system that would have a dynamic range suitable for recording all those changes. Smaller field changes require a higher gain that usually leads to system saturation by larger field changes. On the other hand, a lower gain needed to keep the larger field changes on scale would render the smaller field changes unresolved. The larger field changes are usually relatively slow, varying on timescales of the order of milliseconds and longer (e.g. electric field changes produced by long continuing currents), while the smaller field changes of interest are usually microsecond-scale pulses. One way to enable a field measuring system to record relatively small and relatively short pulses is to allow the larger and slower field changes to decay with a relatively short time constant. In order to avoid distortion of the pulses, this time constant should be much longer than the expected duration of the pulses. As discussed above, time constants satisfying the latter requirement are shorter than a millisecond or so. In this case, some associated field changes varying on a millisecond timescale (e.g. overall field changes produced by K- and M-processes) will be distorted. Specifically, ramp-like electric field changes due to lightning K-processes can be converted to pulses, with the falling edge of the pulse being due to instrumental decay, as opposed to occurring in response to source variation.

In principle, it is possible to compensate for instrumental decay in the post-processing of measured field waveforms, to remove the distortion and reconstruct the undistorted waveform. Also, sometimes there is a need to compare field waveforms recorded by systems characterized by different decay time constants. In this latter case, it is desirable to partially compensate for the shorter instrumental decay to make it equal to that of the other system. Rubinstein et al. (2012) derived the transfer function which can be used for partial compensation of instrumental decay in measured field waveforms and showed how it can be extended to the case of total compensation (reconstruction of undistorted waveforms). Their conversion equation that changes a signal $V(t)$ resulting from integration with a given time constant τ to the signal $V_{new}(t)$, that would have been obtained if a different time constant τ_{new} had been used, is as follows:

$$V_{new}(t) = k_a \left[V(t) + \frac{\tau_{new} - \tau}{\tau_{new}\tau} \int_0^t V(t') \exp\left(-\frac{t - t'}{\tau_{new}}\right) dt' \right] \tag{7.2}$$

where k_a is a constant that can be used to adjust the amplitude of the compensated waveform. The equation can be readily implemented in MATLAB or similar mathematics software. Increasing, in effect, the time constant to 10 s can probably be viewed as the essentially total compensation. It is important to note that in practice the late parts of the signals that have decayed to levels comparable to the noise may be not accurately recoverable.

The open-circuit voltage in the Thevenin equivalent circuit is sometimes written as $V = h_{eff}E$, where $h_{eff} = \varepsilon_0 A/C_a$ (which has the dimension of length) is referred to as the effective height of the antenna. It is difficult to assign any height to a flat-plate antenna that is placed flush with the ground, but it does have some meaning, for example, for a vertical rod or whip antenna, in which case h_{eff} is taken to be one half of the physical height.

Placement of a flat-plate antenna flush with the ground ensures that the electric field to be measured is not influenced by the antenna. This gives an advantage of theoretical calibration of the measuring system (see Eq. 7.1). Any antenna elevated above the ground surface will enhance the field that would exist at the same location in the absence of the antenna. As a result, experimental calibration is required to determine the field enhancement factor (except for the spherical antenna with isolated cutouts, for which the enhancement factor is known; it is equal to 3) the inverse of which is to be used as a multiplier in Eq. 7.1.

Calibration can be done by placing the antenna in a uniform field of a large parallel-plate capacitor or by comparing the antenna output with that of a flush-mounted reference antenna. When calibration is done experimentally, an antenna of any geometry (e.g. a vertical rod (monopole) with or without capacitive loading at its top or an inverted antenna with a grounded "bowl" above the elevated sensing plate) can be used. However, slender antennas are generally not used for measuring fields at short distances from the lightning channel. Such antennas can enhance the electric field to a degree that corona discharge occurs from the antenna. It is impossible to accurately measure electric fields in the presence of corona from the antenna, since, besides the current charging the antenna, there will be corona current transporting charges into the air surrounding the antenna, both currents flowing through the same integrating capacitor across which the output voltage is measured.

If a flat-plate antenna is installed essentially flush with the surface of the roof of a building or other structure, another field enhancement factor, due to the presence of the building, is to be taken into account. This latter enhancement factor can be calculated numerically. For example, Baba and Rakov (2007a), who used the 3D finite-difference time-domain (FDTD) method, estimated that for a building having a plan area of $40 \times 40 \text{ m}^2$ and a height of 20 m the electric field enhancement factor (at the center point of its flat roof) is 1.5, and it is 3.0 if the height of the building is 100 m. For comparison, the enhancement factor on the top of hemispherical structure is independent of its size and equal to 3. The magnitude of vertical electric field at ground level in the immediate vicinity of the building is reduced relative to the case of no building, with this shielding effect becoming negligible at horizontal distances from the building exceeding twice the height of the building. In contrast to the electric field, the magnitude of the magnetic field was found to be not much influenced by the presence of a building. Note that Baba and Rakov (2007a, Table VI) showed that the electric field enhancement due to the presence of building is only slightly influenced by building conductivity ranging from 1 mS/m (dry concrete) to infinity.

The use of long horizontal coaxial cables between the antenna and the associated electronics should be avoided, since the horizontal component of electric field (present due to the finite ground conductivity) can induce unwanted voltages in these cables. The horizontal electric field waveshape is similar to that of the derivative of the vertical field. As a result, the measured field waveform may be a superposition of the vertical field, which is being measured, and the unwanted horizontal field, which causes a distortion of the vertical field waveform by making peaks and valleys sharper than they actually are in the vertical field (Uman, 1987). The problem can be solved by using a fiber-optic link instead of the coaxial cable. Further, significant reduction of noise can be achieved by digitizing signals at the antenna location and digitally transmitting them to recorder.

One can check if the electric field measuring system is working properly by comparing electric field waveforms produced by individual lightning events (e.g. return strokes) with the corresponding magnetic field waveforms. At large distances (>50 km or so), those waveforms are dominated by their radiation components and, hence, their shapes should be identical. Further, the ratios of electric and magnetic field peaks at large distances for sources near ground (return strokes) should be equal to the speed of light (E/B = c). For

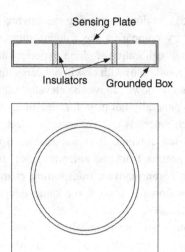

Sensing Plate

Insulators   Grounded Box

Fig. 7.4   Flat-plate antenna for measuring electric field (vertical cross-section and plan views) suitable for flush-mounting in the earth. Drawing by Potao Sun.

elevated vertical sources (see Fig. 5.2), $E/B = c \times \cos\alpha$, where α is the elevation angle measured between the line of sight to the source and ground.

If in Figs. 7.1b, 7.2, and 7.3 the integrating capacitor C is replaced with the resistor R (such that $R \ll 1/(\omega C_a)$ and $R \ll 1/(\omega C_{in})$), the output voltage is proportional to dE/dt. Measured dE/dt waveforms can be numerically integrated over time to obtain E waveforms, although the integration interval should not be too long in order to avoid accumulation of significant error.

An example of flat-plate antenna suitable for flush-mounting in the earth is shown in Fig. 7.4. This antenna has a circular sensing plate mounted, using insulators, in the center of a shallow rectangular box (housing) with a circular opening of a slightly larger radius than that of the sensing plate in its top side. The sensing plate is flush with the top side of the housing, and the latter is grounded. The voltage is measured between the sensing plate and the housing. Such flat-plate antennas are used for measuring E and dE/dt at the ICLRT (both at the Camp Blanding facility and at the Lightning Observatory in Gainesville, Florida). (ICLRT stands for the International Center for Lightning Research and Testing.)

7.1.2 Field mills

If one needs to measure steady or slowly-varying electric fields (on timescales of the order of seconds or more), such as the fair-weather field or fields associated with cloud-charging processes, the condition of $\Delta t \ll \tau$ is difficult or impossible to satisfy. Indeed, in the limit, when E = const, the short-circuit current in the Norton equivalent circuit, which is proportional to dE/dt, is equal to zero. In order to solve this problem, a slowly-varying electric field can be converted into an amplitude-modulated sinusoidal signal using a device which

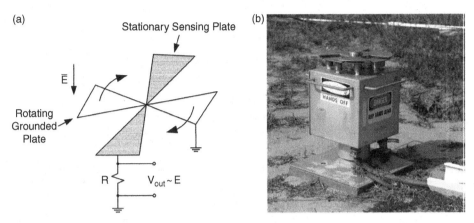

(a) Illustration of the principle of operation of the electric field mill. Adapted from Bazelyan and Raizer (2000). (b) Electric field mill operating at the ICLRT at Camp Blanding, Florida. Photograph by Joseph Howard.

is referred to as an "electrostatic fluxmeter" or "field mill." In this device, the field sensing plate, which is stationary, is periodically shielded and unshielded by the moving grounded plate above it, as schematically shown in Fig. 7.5a. Besides the rotating grounded plate, other ways of modulating the field are possible (for example, a reciprocating shutter or a grounded string vibrating above the sensing plate), but the rotating variety remains the most common one. The charge induced on the sensing plate flows to ground through a resistor when the sensing plate is shielded (covered by the grounded plate) and flows back to the sensing plate when it is uncovered. The voltage across the resistor varies sinusoidall with amplitude proportional to the ambient electric field. The time resolution of field mills is typically a fraction of a second. A photograph of field mill with rotating grounded plate and segmented sensing plate below it, which is operating at the ICLRT at Camp Blanding, Florida, is shown in Fig. 7.5b. The so-called inverted configurations, in which the sensor is elevated and is facing ground, are also used.

7.1.3 Pockels sensors

When some types of crystals – for example, quartz – are placed in an external electric field, the polarization of light (the direction of the light's electric field intensity vector) passing through the crystal changes as a function of field magnitude and the length of the path the light traverses, as schematically shown in Fig. 7.6. This is the electro-optic effect discovered by and named after F. Pockels. It follows that electric fields can be measured by transmitting light through such a crystal and comparing the polarization of the output and input light signals. Electric field sensors operating on this principle are referred to as Pockels sensors. Pockels sensors were used in laboratory spark studies for measurement of the electric field at the tip of leaders and in the leader channels, as well as for measuring electric fields in the immediate vicinity of the triggered-lightning channel.

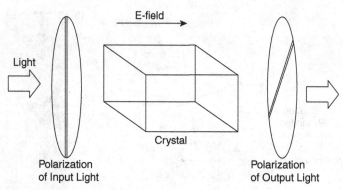

Fig. 7.6   Illustration of Pockels effect. Adapted from Miki et al. (2002).

Pockels sensors have the following advantages for the measurement of electric fields very close to the lightning channel relative to traditional electric field sensors:

1. Pockels sensors usually have no conductive parts and are electrically isolated from the ground. Thus, there is relatively little disturbance of the measured field by the sensor, and such sensors can be placed very close to an electrical discharge or even inside the discharge channel.
2. Crystals used in Pockels sensors can respond to changes in the electric field in a wide frequency range from dc to some gigahertz. However, in many cases the bandwidth of the Pockels-sensor field measuring system (including, besides the Pockels sensor, a light source, a fiber-optic link, and an optical-to-electrical converter) is limited by other elements of the system.
3. Pockels sensors do not contain electronic circuits or power supplies. Thus, Pockels-sensor measurements are less influenced by the unintended coupling of lightning electric and magnetic fields to the measuring system.

Since the dielectric constants of the crystal used in a Pockels sensor and of the crystal holder are different from that of the air, the Pockels sensor will cause some distortion of the electric field. This effect can be accounted for (at least in part) by laboratory calibration of the measuring system. Another potential problem that is more difficult to correct is the distortion of the field due to the attachment of charged particles to the surface of the Pockels sensor when measurements are performed in the lightning source region.

Shown in Fig. 7.7 are two identical systems that were used for measuring the vertical and horizontal components of electric field at distances as short as 0.1 to 1.6 m from the triggered-lightning attachment point by Miki et al. (2002). Each system included a Pockels sensor connected via a fiber-optic link to a light source and an optical-to-electrical (O/E) converter. Each Pockels sensor contained a crystal of KH2PO4 (potassium dihydrogen phosphate also referred to as KDP). The dielectric constant of KDP is about 20 along the optical axis and about 50 in the direction perpendicular to the optical axis. The systems had a relatively large dynamic range, from 20 kV/m to 5 MV/m, but a relatively narrow frequency range, from 50 Hz to 1 MHz. In each Pockels sensor, the crystal was installed

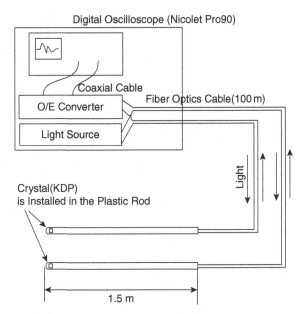

Digital Oscilloscope (Nicolet Pro90)

Coaxial Cable

O/E Converter

Fiber Optics Cable(100 m)

Light Source

Light

Crystal(KDP)
is Installed in the Plastic Rod

1.5 m

Fig. 7.7 Schematic representation of experimental setup used to measure two components of electric field 0.1–1.6 m from the lightning channel with Pockels sensors. Adapted from Miki et al. (2002).

inside a holder made of a dielectric material (fiberglass reinforced plastic) for protection of the crystal from the humidity. Polarized light from the light source was transmitted through the crystal, passed through an analyzer that converts changes in the light polarization to changes in light intensity (included in the O/E converter in Fig. 7.7), converted to an electrical signal, and then digitized and recorded by a digitizing oscilloscope.

Measured electric field waveforms appeared as pulses, with the leading edge of the pulse being due to the leader and the trailing edge due to the return stroke. Vertical electric field pulse peaks were in the range from 176 kV/m to 1.5 MV/m (the median is 577 kV/m), and horizontal electric field pulse peaks were in the range from 495 kV/m to 1.2 MV/m (the median is 821 kV/m).

7.2 Wideband magnetic field measurements

To measure the magnetic field produced by lightning processes a loop of wire can be used as an antenna. According to Faraday's Law, a time-varying magnetic field passing through an open-circuited loop of wire will induce a voltage (electromotive force) at the terminals of the loop (see Fig. 7.8). The induced voltage is proportional to the rate of change of magnetic flux passing through the loop area. Assuming that the loop area, A, is small enough to consider the normal component of magnetic flux density, $B_n = B \cos\alpha$, where α is the angle

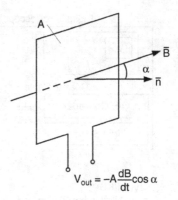

$$V_{out} = -A\frac{dB}{dt}\cos\alpha$$

Fig. 7.8 Illustration of the principle of operation of the loop antenna. Drawing by Potao Sun.

between the magnetic flux density vector and the normal to the plane of the loop, to be constant over that area, we can express the magnitude of induced voltage as follows:

$$V = A\frac{dB_n}{dt} \tag{7.3}$$

When $\cos\alpha = 1$ ($\alpha = 0$), the induced voltage is maximum, and when $\cos\alpha = 0$ ($\alpha = 90°$), the induced voltage is zero. It follows that a vertical loop antenna in a fixed position is directional in that the magnitude of voltage induced across its terminals is a function of the direction to the source, and two such antennas with orthogonal planes can be used for magnetic direction finding (see Chapter 8). In order to obtain the horizontal (azimuthal) component of magnetic field, which is the dominant component for essentially vertical lightning channels, two vertical loop antennas are required, unless the direction to the lightning channel is known (for example, in the case of rocket-triggered lightning). Since the signal at the output of a loop antenna is proportional to the magnetic field derivative, the signal must be integrated to obtain the field. This can be accomplished using either an RC or RL circuit, or the measured field derivative signal can be integrated numerically. We will consider below the case of RC integrator.

In the following, we will assume that B is normal to the plane of the loop antenna ($\alpha = 0$), so that $B = B_n$. The voltage induced at the terminals of a loop antenna is the open-circuit voltage, $Aj\omega B$, and hence it can be used for building the Thevenin equivalent circuit of the antenna. The source impedance is predominantly inductive, $j\omega L$. The overall equivalent circuit including, besides the antenna, the RC integrator and input impedance (input resistance in parallel with input capacitance) of the recorder is shown in Fig. 7.9.

In contrast with the electric field antenna (see Fig. 7.2), the integrating capacitor in Fig. 7.9 has two discharge paths, one through the input resistance of the recorder (similar to Fig. 7.2) and the other through resistor R of the integrating circuit and the source (the ideal voltage source has zero impedance). As a result, there are three conditions for undistorted recording of magnetic field with the measuring system shown in Fig. 7.9. The first one, $R \gg 1/\omega C$ ($\omega \gg 1/(RC)$; C_{in} is neglected), determines the lower frequency limit and is equivalent to the $\Delta t \ll \tau$ ($\tau = RC$) condition. The second one, $R \gg \omega L$ ($\omega \ll R/L$),

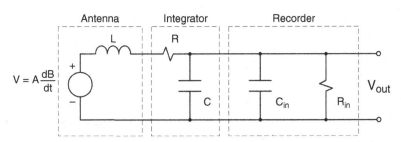

Fig. 7.9   Thevenin equivalent circuit of magnetic field antenna shown along with the integrating circuit and the input impedance of recorder. Drawing by Potao Sun.

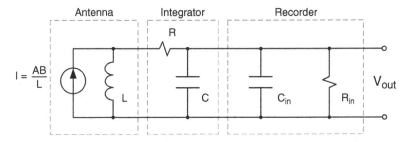

Fig. 7.10   Norton equivalent circuit of magnetic field antenna shown along with the integrating circuit and the input impedance of recorder. Drawing by Potao Sun.

determines the upper frequency limit. The third one, $R_{in} \gg R$, requires that C discharges primarily through R, not R_{in}. Under those three conditions, the output voltage is independent of frequency and given by

$$V_{out} = \frac{AB}{RC} \tag{7.4}$$

The Norton equivalent circuit can be obtained from the Thevenin equivalent circuit by finding the antenna short-circuit current, $I = Aj\omega B/j\omega L = AB/L$, which is independent of frequency, and placing it (as an ideal current source) in parallel with the antenna impedance, $j\omega L$. The overall circuit, including the integrator and recorder, is shown in Fig. 7.10. The choice between the Thevenin and Norton equivalent circuits is a matter of taste.

Magnetic field measuring circuits are rarely passive; active integrators and amplifiers are usually required. A loop antenna developed by George Schnetzer and its Thevenin equivalent circuit are shown in Fig. 7.11. The antenna is made of 50 Ω coaxial cable, each end of which is terminated in 50 Ω resistance. The inner conductor of the cable is the actual wire comprising the loop antenna. The grounded outer shield is necessary for shielding the inner conductor from electric fields. The shield is broken at the top of the antenna, so that no circulating currents can flow in the shield. The output voltage is taken from only one end of the cable and

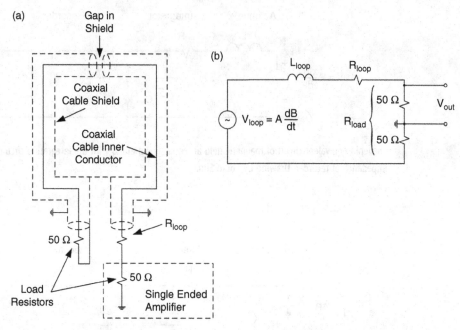

Fig. 7.11 Loop antenna with single-ended output developed by George Schnetzer. (a) Schematic diagram; (b) Thevenin equivalent circuit. Drawing by Potao Sun.

the 50 Ω termination on that end of the cable takes the form of the input resistance of an amplifier that the antenna is connected to. The other end of the cable is terminated by soldering a 50 Ω resistor between the inner conductor and the outer shield. The magnitude of output voltage of the antenna is proportional to dB/dt (should be integrated to obtain B) and is given, for the case of $(R_{loop} + R_{load}) \gg \omega L$ or $\omega \ll (R_{loop} + R_{load})/L$, by

$$V_{out} = \frac{R_{load}A}{2(R_{loop} + R_{load})} \frac{dB}{dt} \tag{7.5}$$

where A is the area of the antenna, $R_{load} = 100\ \Omega$ (two 50 Ω resistors in series), and R_{loop} is the total resistance of the loop antenna including its inherent resistance (very small and can be neglected) and any externally added resistance which affects both the gain and the bandwidth of the antenna. The −3-dB upper limit f_0 of the frequency bandwidth of the antenna is given by the following equation:

$$f_0 = \frac{R_{loop} + R_{load}}{2\pi L_{loop}} \tag{7.6}$$

where L_{loop} is the inductance of the antenna. For example, $f_0 = 60$ MHz for $L_{loop} = 4\ \mu H$ and $R_{loop} = 1410\ \Omega$.

The loop antenna shown in Fig. 7.11 is used at the ICLRT (both at the Camp Blanding facility and at the Lightning Observatory in Gainesville, Florida).

7.3 Characteristics of wideband electric and magnetic fields produced by lightning

In designing field measuring systems, one needs to know expected magnitudes and durations of signals to be recorded. Different lightning processes produce different electromagnetic signatures, these signatures change with distance, and at the same distance there is large variation in source strength. Both variations in the source and with distance should be considered. Given in this section is a summary of characteristics of lightning electric and magnetic fields expected at different distances from the source.

Characteristics of microsecond-scale electric field pulses (including pulse durations and interpulse intervals) associated with various lightning processes are summarized in Table 7.1. Continuing currents have durations ranging from a few to some hundreds of milliseconds. The overall flash duration is typically 200 ms or so, but can exceed 1 s.

At ground level and at distances greater than a few kilometers, the initial electric field peak is dominated by its radiation component. Typical electric field peak values normalized to 100 km are about 6 V/m and 3 V/m for negative first and negative subsequent return strokes, respectively. The largest radiation field peaks due to stepped and dart-stepped leaders are typically a factor of 10 smaller than the corresponding return-stroke field peak. Radiation fields vary inversely with distance (1/r dependence), if propagation effects due to finite ground conductivity (Section 5.6) can be neglected. Generally, the typical radiation field peak values normalized to 100 km can be scaled to either smaller or larger distances in the range from about 5 to about 200 km. For example, if the field peak at 100 km is 6 V/m, it is expected to be 60 V/m at 10 km and 3 V/m at 200 km. The corresponding magnetic radiation field peaks can be readily found by dividing the electric field peak by the speed of light (3×10^8 m/s) to find the magnetic flux density (B) and by the intrinsic impedance of free space (377 Ω) to find the magnetic field intensity (H). At a given distance, the field can be at least a factor of five greater and a factor of five smaller, due to variation in the source.

At very close distances (tens to hundreds of meters) the leader and return-stroke electric filed changes are essentially electrostatic and approximately equal to each other (at least for rocket-triggered-lightning strokes). They each vary as approximately 1/r. For triggered-lightning strokes, which are similar to subsequent strokes in natural lightning, the magnitudes of electric field at 15 and 30 m are typically about 100 and 50 kV/m, respectively. The half-peak width of the V-shaped pulse formed by the leader and return-stroke electric filed changes increases with distance and is about 2 and 4 μs at 15 and 30 m, respectively. At 500 m, the magnitude of electric field is typically about 2 kV/m and the half-peak width is about 100 μs. The magnetic field at very close distances varies with distance as 1/r. For triggered-lightning strokes, the magnitudes of magnetic flux density pulses at 15 and 30 m are typically about 200 and 100 μT, respectively. The corresponding typical half-peak widths are 10 to 15 μs. Additional information on very close electric and magnetic field waveforms is found in works of Rubinstein et al. (1995), Crawford et al. (2001), and Schoene et al. (2003a).

Table 7.1 Characterization of microsecond-scale electric field pulses associated with various lightning processes. Adapted from Rakov and Uman (2003)

Type of Pulses	Dominant Polarity		Typical Total Pulse Duration, μs	Typical Time Interval Between Pulses, μs	Comments
	Atmospheric Electricity Sign Convention	Physics Sign Convention			
Return stroke in negative ground flashes	Positive	Negative	30–90 (zero-crossing time)	60 × 10³	3–5 pulses per flash
Stepped leader in negative ground flashes	Positive	Negative	1–2	15–25	Within 200 μs just prior to a return stroke
Dart-stepped leader in negative ground flashes	Positive	Negative	1–2	6–8	Within 200 μs just prior to a return stroke
Initial breakdown in negative ground flashes	Positive	Negative	20–40	70–130	Some milliseconds to some tens of milliseconds before the first return stroke
Initial breakdown in cloud flashes	Negative	Positive	50–80	600–800	The largest pulses in a flash
Regular pulse burst in both cloud and negative ground flashes	Both polarities are about equally probable		1–2	5–7	Occur later in a flash; 20–40 pulses per burst
Narrow bipolar pulse	Negative	Positive	10–30	–	Produced by compact intracloud discharges (CIDs)

Notes:

1. Polarity refers to polarity of the initial half cycle in the case of bipolar pulses.

2. Typical values are based on a comprehensive literature search and unpublished experimental data acquired by the University of Florida Lightning Research Group.

3. In the atmospheric electricity sign convention, a downward-directed electric field or electric field change vector is assumed to be positive and it is negative in the physics sign convention.

The overall (essentially electrostatic) field change due to a lightning stroke at 100 km is of the order of 1 V/m and it is of the order of 10^4 V/m (10 kV/m) at 5 km (Uman, 1987). The electrostatic field at relatively large distances varies as $1/r^3$.

If the expected field peak at close or far distance is to be estimated from the return-stroke peak current, the approximate equations based on the transmission-line model (see Table 5.2) can be used.

The quasi-static electric field at ground level under fair-weather conditions is about 100 V/m (directed downward) and can increase to 1–10 kV/m (usually directed upward) during thunderstorms. The latter values are limited by a blanket of space charge (usually of positive polarity) in the air near Earth's surface produced by corona from various pointed objects on the ground. Corona space charge creates its own electric field that opposes the external field that would be measured at ground in the absence of corona.

7.4 Summary

A sensor that is commonly used to measure the lightning vertical electric field is a metallic disk placed flush with the ground surface, the so-called flat-plate antenna. This configuration ensures that the electric field to be measured is not influenced by the antenna. This gives an advantage of theoretical calibration of the measuring system. Any antenna elevated above the ground surface will enhance the field that would exist at the same location in the absence of antenna. As a result, experimental calibration is generally required to determine the field enhancement factor. The output signal of electric field antenna is proportional to dE/dt and hence should be integrated to obtain E. To measure the magnetic field produced by lightning processes a loop of wire can be used as an antenna. In order to obtain the horizontal (azimuthal) component of magnetic field, which is the dominant component for essentially vertical lightning channels, two vertical loop antennas are required, unless the direction to the lightning channel is known (for example, in the case of rocket-triggered lightning). The output signal of magnetic field antenna is proportional to dB/dt and hence should be integrated to obtain B. When measurements are performed on the roof of a building, the electric field can be significantly enhanced, while the magnetic field is not much influenced by the presence of the building. The magnitude of vertical electric field at ground level in the immediate vicinity of the building is reduced relative to the case of no building, with this shielding effect becoming negligible at horizontal distances from the building exceeding twice the height of the building. Antennas can be represented by either Norton or Thevenin equivalent circuits. The undesired instrumental decay in measured electric field waveforms can be compensated for, so that the instrumental distortion is in effect removed. A slowly-varying electric field can be measured by converting it into an amplitude-modulated sinusoidal signal using an electrostatic fluxmeter (field mill). Electric fields in the immediate vicinity of the lightning channel can be measured by transmitting light through a crystal that exhibits Pockels effect and comparing the polarization of the output and input light signals.

Questions and problems

7.1. Define the instrumental decay time constant. What is the condition for measuring the electric field without instrumental distortion in (a) the time domain and (b) the frequency domain?

7.2. What is the difference between the "fast antenna" and "slow antenna" electric field measuring systems?

7.3. Why does an elevated flat-plate antenna result in higher measured electric fields than a similar antenna mounted flush with the ground?

7.4. Describe the principle of operation of the electric field mill.

7.5. What is the Pockels sensor?

7.6. Why does the signal at the output terminals of a vertical loop antenna depend on the direction to the source? Where is the source, if the two identical loop antennas whose planes are oriented in the same direction recorded signals of the same magnitude but opposite polarity?

7.7. Suppose that the peak electric field of a lightning return stroke measured at a distance of 100 km is 10 V/m. What are the corresponding (a) peak magnetic flux density and (b) peak magnetic field intensity?

7.8. Suppose you need to measure the electric field produced by dart leaders in triggered lightning at a distance of 30 m from the channel (see Fig. 4.37) with a 0.2 m^2 flat-plate antenna and a digitizing oscilloscope that are connected by a fiber-optic link. Assume that the fiber-optic link transmitter has an input resistance of 1 MΩ, an input capacitance of 30 pF, and three input voltage settings: ± 0.1 V, ± 1 V, and ± 10 V. Can you use a passive integrator at the antenna for this measurement or an active integrator is required?

Further reading

Cooray, V., ed. (2012). *Lightning Electromagnetics*, 950 pp., London: IET.

Uman, M. A. (1987). *The Lightning Discharge*, 377 pp., San Diego, California: Academic Press.

Uman, M. A. (2001). *The Lightning Discharge*, 377 pp., Mineola, New York: Dover.

8 Electromagnetic methods of lightning location

8.1 Introduction

There are many individual physical processes in cloud and ground lightning flashes. Each of these processes has associated with it electric and magnetic fields. Lightning is known to emit significant electromagnetic energy in the frequency range from below 1 Hz to near 300 MHz, with a peak in the frequency spectrum near 5–10 kHz for lightning at distances beyond 50 km or so. Further, electromagnetic radiation from lightning is detectable at even higher frequencies, for example, in the microwave, 300 MHz to 300 GHz, and, obviously, in visible light, roughly 10^{14}–10^{15} Hz. At frequencies higher than those of the spectrum peak, the field spectral amplitude varies roughly inversely proportionally to the frequency up to 10 GHz or so (Pierce, 1977). Also, lightning is known to produce X-rays (up to 10^{20} Hz or more), although, at ground level, they are usually not detectable beyond a kilometer or so from the source. In general, any observable electromagnetic signal from a lightning source can be used to detect and locate the lightning process that produced it. Only radio-frequency methods are considered in detail here.

8.2 Principles of lightning location

8.2.1 General

The three most common electromagnetic radio-frequency-locating techniques include the magnetic direction finding (MDF), time of arrival (TOA), and interferometry. For each of these techniques, the type of locating information obtained depends on the frequency f (or equivalently on the wavelength $\lambda = c/f$, where c is the speed of light) of the radiation detected (Rakov and Uman, 2003, Ch. 17). For detected signals whose wavelengths are very short compared to the length of a radiating lightning channel, for example, the very-high-frequency (VHF) range where f = 30–300 MHz and λ = 10–1 m, the whole lightning channel can, in principle, be imaged in three dimensions. For wavelengths that exceed or are a significant fraction of the lightning channel length, for example, the very-low-frequency (VLF) range where f = 3–30 kHz and λ = 100–10 km and the low-frequency (LF) range where f = 30–300 kHz and λ = 10–1 km, generally, only a small number of locations can be usefully obtained. In the case of single location for a cloud-to-ground discharge it is usually interpreted as some approximation to the ground strike point. The mechanisms of radiation in the high-frequency (HF) region of the spectrum and above are not fully understood. It is thought that this radiation (particularly VHF, 30–300 MHz) is caused by numerous small sparks occurring during the formation of new channels; that is, by the electrical breakdown of air rather than by high-current pulses propagating in preexisting channels.

Accurate lightning locating systems, whether they image the whole lightning channel or locate only the ground strike points or the cloud charge centers, necessarily employ multiple sensors. Single-station ground-based sensors, such as the lightning flash counters, detect the occurrence of lightning, but cannot be used to locate it on an individual flash basis; nor are they designed to do so, because of the wide range of amplitudes and waveshapes associated with individual events. Nevertheless, with single-station sensors, one can assign groups of flashes to rough distance ranges if data are accumulated and "averaged" for some period of time. There are many relatively simple commercially available single-station devices that purport to locate lightning. Most operate like AM radios with the amplitude of the radio static being used to gauge the distance to the individual lightning flashes, a technique inherently characterized by large errors.

Single-station optical sensors on Earth-orbiting satellites detect the light scattered by the volume of cloud that produces the lightning and hence cannot locate to an accuracy better than about 10 km, about the diameter of a small cloud. Additionally, satellite-based sensors cannot distinguish between cloud and ground discharges. The next-generation series of Geostationary Operational Environmental Satellite (GOES-R) is planned to carry a Geostationary Lightning Mapper (GLM), which will monitor lightning continuously over a wide field of view.

In the following, we will discuss how individual sensors measuring various properties of the lightning electromagnetic radiation can be combined into systems to provide lightning locations. More details can be found in the reviews by Cummins and Murphy (2009), Rakov (2013), and Nag et al. (2015) and in references therein.

As noted above, there are three major lightning locating methods: magnetic direction finding (MDF), time-of-arrival (TOA), and interferometry. The MDF and long-baseline (hundreds of kilometers) TOA systems usually operate in the VLF/LF range and report one location per lightning event. The interferometry and shorter-baseline (tens of kilometers or less) TOA systems usually operate at VHF and provide multiple locations per lightning event (VHF images of lightning channels). The MDF, TOA, and interferometric methods are reviewed in Sections 8.2.2, 8.2.3, and 8.2.4, respectively. Lightning locating systems operating on global scale utilize methods that are capable of extracting source information from electromagnetic signals dominated by ionospheric reflections. These latter methods are considered in Section 8.2.5.

8.2.2 Magnetic Direction Finding (MDF)

Two vertical and orthogonal loops with planes oriented N–S and E–W, each measuring the magnetic field from a given vertical radiator, can be used to obtain the direction to the

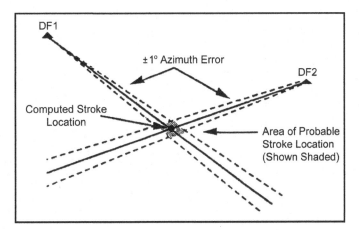

Fig. 8.1 Determination of lightning stroke location when only two direction finders (DFs) detect the stroke. The solid lines represent the measured azimuths to the stroke; the broken lines represent the ±1° angular random error in the azimuth measurements. The solid circle indicates the computed stroke location; the shaded region indicates the uncertainty in location of the stroke. Adapted from Holle and Lopez (1993).

source. This is the case because the output voltage of a given loop, by Faraday's law, is proportional to the cosine of the angle between the magnetic field vector and the normal vector to the plane of the loop (Chapter 7). For a vertical radiator, the magnetic field lines are circles coaxial with respect to the source. Hence, for example, the loop whose plane is oriented N–S receives a maximum signal if the source is north or south of the antenna, while the orthogonal E–W loop receives no signal. In general, the ratio of the two signals from the loops is proportional to the tangent of the angle between north and the source as viewed from the antenna. Thus a pair of two orthogonal loops can be used as a direction finder (DF). As illustrated in Fig. 8.1 for a two-DF system, the intersection of two direction (azimuth) vectors from the DF to the apparent source, provides a stroke location, but a location containing error because each azimuth vector has some random angular error and may have some systematic error. If a three-DF system is employed, each pair of DFs provides a location, so there are three locations, the distance between the locations providing some measure of the system error, as illustrated in Fig. 8.2. For three or more DF responses to a lightning return stroke, the optimal estimate of the location is best found using a χ^2 minimization technique.

Crossed-loop magnetic direction finders (DFs) used for lightning detection can be divided into two general types: narrow-band (tuned) DFs and gated wideband DFs. In both cases, the direction-finding technique involves an implicit assumption that the radiated electric field is oriented vertically and the associated magnetic field is oriented horizontally and perpendicular to the propagation path.

Narrow-band DFs have been used to detect distant lightning since the 1920s (Horner, 1954, 1957). They generally operate in a narrow frequency band with the center frequency in the 5–10 kHz range, where attenuation in the Earth-ionosphere waveguide is relatively low and where the lightning signal energy is relatively high. Before the development of

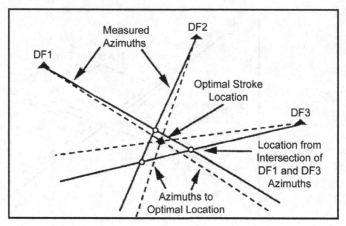

Determination of lightning stroke location when three DFs detect the stroke. The solid lines represent the measured azimuths to the stroke. The open circles indicate the three possible locations defined by the three different intersections of the azimuth vectors. The optimal stroke location (solid circle) is determined by minimizing the χ^2 function. The broken lines show the azimuth vectors to the computed optimal location. Adapted from Holle and Lopez (1993).

weather radars in the 1940s, lightning locating systems were the primary means of identifying and mapping thunderstorms at medium and long ranges.

A major disadvantage of narrow-band DFs is that for lightning at distances less than about 200 km, those DFs have inherent azimuthal errors, called polarization errors, of the order of 10° (Nishino et al., 1973; Kidder, 1973). These errors are caused by the detection of magnetic field components from non-vertical channel sections, whose magnetic field lines form circles in a plane perpendicular to the non-vertical channel section, and by ionospheric reflections, sky waves, whose magnetic fields are also improperly oriented for direction finding of the ground strike point.

To overcome the problem of large polarization errors at short ranges inherent in the operation of narrow-band DFs, gated wideband DFs were developed in the early 1970s (Krider et al., 1976). Direction finding is accomplished by sampling (gating on) the N–S and E–W components of the initial peak of the return-stroke magnetic field, that peak being radiated from the bottom hundred meters or so of the channel in the first microseconds of the return stroke. Since the bottom of the channel tends to be straight and vertical (perpendicular to the ground), the magnetic field lines form circles in a horizontal plane. Additionally, a gated DF does not record ionospheric reflections since those reflections arrive long after the initial peak magnetic field is sampled. The operating bandwidth of the gated wideband DF is typically from a few kilohertz to about 500 kHz. Interestingly, although an upper frequency response of many megahertz is needed to assure accurate reproduction of the incoming radiation field peak, particularly if the propagation is over salt water, practical DFs only need an upper frequency response of a few hundred kilohertz in order to obtain an azimuthal error of about 1°. This is the case because the ratio of the peak signals in the two loops is insensitive to the identical distortion produced by the identical associated electronic circuits of the two loops. Thus, the gated wideband DF can operate at frequencies below the AM radio band and below the frequencies of some aircraft navigational transmitters, either of which could otherwise cause unwanted directional noise.

Gated wideband DFs, as well as narrow-band DFs, are susceptible to site errors. Site errors are a systematic function of direction but generally are time-invariant. These errors are caused by the presence of unwanted magnetic fields due to non-flat terrain and to nearby conducting objects, such as underground and overhead power lines and structures, being excited to radiate by the incoming lightning fields. In order to eliminate site errors completely, the area surrounding a DF must be flat and uniform, without significant conducting objects, including buried ones, nearby. These requirements are usually difficult to satisfy, so it is often easier to measure the DF site errors and to compensate for any that are found than to find a location characterized by tolerably small site errors. Once corrections are made, the residual errors have been reported (using independent optical data) to be usually less than two to three degrees (e.g. Mach et al., 1986).

Since it is not known a priori whether a stroke to ground lowers positive or negative charge, there is an 180° ambiguity in stroke azimuth from the measurement of only the orthogonal magnetic fields. That ambiguity is resolved in DF systems by the measurement of the associated electric field (Chapter 7) whose polarity indicates the sign of the charge transferred to the ground.

8.2.3 Time-of-Arrival (TOA) technique

A single time-of-arrival sensor provides the time at which some portion of the lightning electromagnetic field signal arrives at the sensing antenna. Time-of-arrival systems for locating lightning can be divided into three general types: (1) very short baseline (tens to hundreds of meters), (2) short baseline (tens of kilometers), and (3) long baseline (hundreds to thousands of kilometers). Very short and short baseline systems generally operate at VHF – that is, at frequencies from 30 to about 300 MHz – while long baseline systems generally operate at VLF and LF, 3–300 kHz. It is generally thought that VHF radiation is associated with air breakdown processes, while VLF signals are due to current flow in conducting lightning channels. Short baseline systems are usually intended to provide images of lightning channels and to study the spatial and temporal development of discharges. Long baseline systems are usually used to identify the ground strike point or the "average" location of the flash.

A very short baseline system (tens to hundreds of meters; e.g. Cianos et al., 1972; Taylor, 1978; Ray et al., 1987) is composed of two or more VHF time-of-arrival (TOA) receivers whose spacing is such that the time difference between the arrival of an individual VHF pulse from lightning at those receivers is short compared to the time interval between pulses, which is some microseconds to hundreds of microseconds. The locus of all source points capable of producing a given time difference between two receivers is, in general, a hyperboloid, but if the receivers are very closely spaced, the hyperboloid degenerates, in the limit, into a plane on which the source is found. Two time differences from three very closely spaced receivers yield two planes whose intersection gives the direction to the source; that is, its azimuth and elevation. To find the source location, as opposed to determination of the direction to the source, two or more sets of three closely spaced receivers, the sets being separated by tens of kilometers or more, must be used. Each set of receivers is basically a TOA direction finder, and the intersection of two or more direction vectors yields the location.

Short-baseline TOA systems (e.g. Proctor, 1971; Lennon and Poehler, 1982; Rison et al., 1999) are typically networks of 5–15 stations that make use of time-of-arrival information for three-dimensional (3D) mapping of lightning channels. A portable version of such system has been developed by researchers at the New Mexico Institute of Mining and Technology. This system is presently referred to as the Lightning Mapping Array (LMA) and has recently become a major tool for both lightning research and operational applications. The short-baseline VHF TOA systems provide electromagnetic images of the developing channels of any type of lightning flash. An example of LMA image of the upward positive leader and initial continuous current of a flash triggered using the rocket-and-wire technique at Camp Blanding, Florida, is shown in Fig. 8.3.

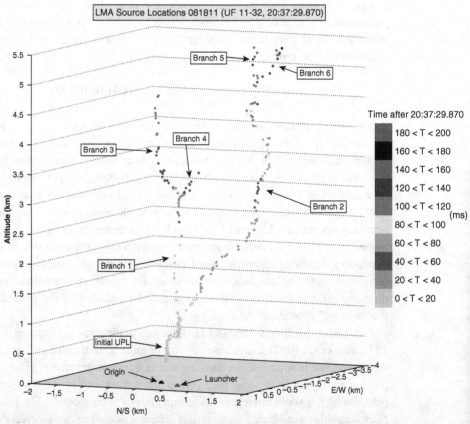

Fig. 8.3   Three-dimensional view of the VHF source locations associated with the upward positive leader and initial continuous current of flash UF 11–32 triggered at Camp Blanding, Florida, on August 18, 2011. The image was obtained with LMA. The sources span 200 ms and are color coded in time in 20 ms time windows according to the key at right. Some sources have been removed. Adapted from Hill et al. (2012).

One of the first long baseline (hundreds to thousands of kilometers) TOA systems operated at VLF/LF (Lewis et al., 1960). It employed a pair of receiving stations in Massachusetts with a bandwidth of 4–45 kHz, separated by over 100 km (the overall network was composed of four stations), to compare differences in the times of arrival of the signals at the two stations and hence to determine directions to the causative lightning discharge in western Europe. The two-station system was basically a direction finder similar to the very short baseline systems described above, but operating at lower frequencies and longer baseline. The resultant "directions" compared favorably with the locations reported by the British Meteorological Office's narrow-band DF network. Spherical geometry was used to account for propagation over the Earth's surface in finding the locus of points for a constant measured arrival time difference between receivers.

Another long-baseline TOA system, called the Lightning Positioning and Tracking System (LPATS), was developed in the 1980s. The LPATS, operating at VLF/LF, used electric field whip antennas at stations 200–400 km apart to determine locations via the measured differences between signal arrival times at the stations. In the frequency band used, return-stroke waveforms were generally the largest and hence most easily identified. In general, responses from four stations (three time differences) are needed to produce a unique stroke location, since the hyperbolas on the Earth's surface from only two time differences can, in general, intersect at two different points, as illustrated in Fig. 8.4. For cloud-to-ground lightning near or within the network, there is often only one solution (see Fig. 8.5), so in this case, the three-station solution suffices.

8.2.4 Interferometry

In addition to radiating isolated pulses, lightning also produces noise-like bursts of electromagnetic radiation lasting for tens to hundreds of microseconds. These bursts are difficult to locate using TOA techniques due to the difficulty in identifying the individual pulses. In the case of interferometry, no identification of individual pulses is needed, since the interferometer measures phase difference between narrow-band signals corresponding to these noise-like bursts received by two or more closely spaced sensors. The simplest lightning interferometer consists of two antennas some meters apart, each antenna being connected via a narrow-band filter to a receiver. The antennas, filters, and receivers are identical. The output signals of the two receivers are sent to a phase detector that produces a voltage that is proportional to the difference in phase between the two quasi-sinusoidal signals. The phase difference defines, as does the time difference in very-short-baseline TOA systems, a plane on which the source is located; that is, one direction angle to the VHF source. To find the azimuth and elevation of a source, three receiving antennas with two orthogonal baselines are needed, at minimum. To locate the source in three dimensions, two or more synchronized interferometers, each effectively acting as a direction finder, separated by a distance of the order of 10 km or more, are needed. The principles of interferometric lightning location are described in detail by Lojou et al. (2009). Examples of interferometric images of lightning processes are found in Figs. 4.6 and 4.12.

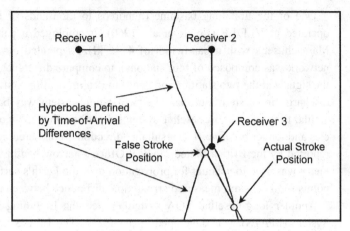

Fig. 8.4 Determination of lightning stroke location by three TOA receivers when the solution is not unique. Shown are two hyperbolas, defined by the TOA differences, that intersect at two points (open circles); one point corresponds to the actual stroke position and the other is a false solution. Adapted from Holle and Lopez (1993).

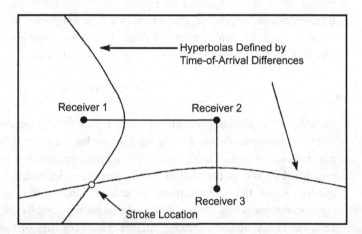

Fig. 8.5 Determination of lightning stroke location by three TOA receivers when the solution is unique. Shown are two hyperbolas, defined by the TOA differences, that intersect to define the unique location of the stroke (open circle). Adapted from Holle and Lopez (1993).

Most interferometric systems operate over very narrow frequency bands (a few hundred kilohertz to a few megahertz in the VHF/UHF bands, where UHF indicates the ultra-high-frequency range, from 300 MHz to 3 GHz), since this allows the system to have high sensitivity in a specific "quiet" band of operation. However, it also makes the system performance subject to local broadband interference, may not provide the highest possible signal-to-noise ratio (SNR), and places a specific limitation in the spacing of the antenna array elements to avoid arrival time (phase) ambiguity. There is a recent trend toward using broadband interferometry (e.g. Morimoto et al., 2004; Stock et al., 2014). This trend is made possible by the advent of affordable broadband RF and digital signal processing electronics.

8.2.5 Lightning location on global scale

Global lightning locating systems employ a relatively small (some tens or less) number of sensors, so that the baselines are of the order of thousands of kilometers and signals are dominated by ionospheric reflections. The VLF range (3–30 KHz) is used. Two location methods employed by currently operating systems are outlined below.

Time-of-Group-Arrival (TOGA) method. This method is based on the fact that lightning VLF signals (sferics) propagating in the Earth-ionosphere waveguide experience dispersion, in that the higher-frequency components arrive earlier than the lower-frequency components (e.g. Dowden et al., 2002). The TOGA, a quantity that can be derived from the measured sferic waveform, is related to the distance traveled by the sferic. The TOGA lightning location method is implemented in the World Wide Lightning Location Network (WWLLN).

MDF and TOA methods combined with a lightning waveform recognition algorithm. Lightning locations are obtained using the MDF and long-baseline TOA methods, similar to those described in Sections 8.2.2 and 8.2.3, respectively, in conjunction with a lightning waveform recognition algorithm. The latter relies on a bank of "canonical" waveforms corresponding to propagation distances of the order of thousands of kilometers (Said et al., 2010). This method is implemented in the Global Lightning Dataset (GLD360).

8.3 Performance characteristics

Generally, a modern VLF–MF (MF indicates the medium frequency range, from 300 kHz to 3 MHz) lightning locating system (LLS) is expected to record, in separate categories, and locate over a certain area all or most cloud-to-ground strokes of either polarity, as well as cloud discharges. Also expected for each discharge is a measure of its intensity, usually in the form of peak current inferred from measured electric or magnetic field. Accordingly, system's performance can be evaluated using the following characteristics:

(a) cloud-to-ground (CG) flash detection efficiency;
(b) CG stroke detection efficiency;
(c) cloud discharge detection efficiency;
(d) percentage of misclassified events (particularly cloud discharges assigned to the positive or negative CG stroke category);
(e) location accuracy (or location error);
(f) peak current estimation error.

Given below are definitions of LLS performance characteristics, followed by a discussion of various approaches to their evaluation.

In general, the detection efficiency is the fraction (usually expressed in percent) of the total events occurred that are detected by the system and is ideally equal to 100 percent. While the CG stroke detection efficiency can be readily defined (since these strokes involve a unique and observable feature – luminous channel to ground – and the total number of occurred events can be practically determined), the cloud discharge detection efficiency concept is rather uncertain. Indeed, there are many cloud discharge processes (some of them poorly understood) occurring on different spatial- and time-scales and apparently exhibiting no unique and readily observable features. As a result, the total number of occurred events is generally unknown. In practice, if all cloud discharge events are accepted as "counts," the number of detected cloud discharges may be largely determined by the local noise level and system's signal transmission rate limit.

In defining the CG flash detection efficiency, probably the most important performance characteristic for lightning locating systems used for the determination of ground flash density, a flash is considered to be detected when at least one stroke of the flash was detected. A similar approach could be applied to cloud flashes, although one would need to decide whether a single "count" constitutes a flash and how to assign multiple "counts" to individual flashes.

The location error is the distance between the actual location and that reported by the system. In general, the location error consists of random and systematic components. The latter in some cases can be accounted for (e.g. site errors in MDF systems).

The peak current estimation error is the difference between the actual peak current value and that reported by the system, usually expressed as a percentage of the actual peak current. Peak currents are estimated by lightning locating systems using either an empirical or model-based field-to-current conversion equation. There are reasonable field-to-current conversion equations for CG strokes, but not for cloud discharge processes.

In order to evaluate the performance characteristics listed above, independent (ground-truth) data are needed. For example, discharges occurring at a precisely known location equipped with a current-measuring device (tall tower or lightning-triggering facility) can be used for estimating the location accuracy and peak current estimation error. Detection efficiencies and percentage of misclassified events are usually estimated based on time-resolved optical recordings. Sometimes, lightning-related damage to various objects (buildings, trees, etc.) is used in estimating location errors, although identification of causative lightning event in this approach is not unique due to insufficient accuracy of timing information (usually not known within better than a minute). Less definitive evaluations of lightning locating system's performance characteristics are possible via modeling or comparison with a more accurate system operating in the same area. As of today, only a limited number of ground-truth studies have been performed, particularly for first strokes in negative CG flashes, positive CG flashes, and cloud discharges.

In some applications (e.g. tracking of thunderstorm cells), the tracking ability may be more important than detection of individual lightning discharges. Performance of the systems intended primarily for such applications is often tested against radar or infrared satellite imagery, with a good correspondence between detected lightning and regions of high radar reflectivity or low cloud-top temperature being viewed as indicative of system's output validity. For early warning, the ability to detect the first lightning is probably the most important performance characteristic.

It is not clear how to define the performance characteristics for VHF lightning channel imaging systems in terms of lightning events, as opposed to individual VHF sources. Limitations in sensitivity usually prevent these systems from directly detecting and mapping positive leaders. Further, supplementary information about return strokes is usually needed to reliably distinguish between cloud flashes (including attempted leaders developing toward the ground) and CG flashes, because the VHF radiation directly associated with return strokes may be limited and difficult to detect. Also, no peak current estimates are possible. Nevertheless, VHF lightning channel imaging systems represent a very valuable tool for studying detailed lightning morphology and evolution, particularly inside the cloud, and are often used in testing other types of lightning locating systems.

8.4 Examples of modern lightning locating systems

One VHF lightning channel imaging system (LMA) and six single location per lightning event networks, three VLF/LF (NLDN, LINET, and USPLN), one ELF–HF (ENTLN), and two VLF (WWLLN and GLD360) systems, are briefly reviewed here as representative examples of modern lightning locating systems. Actually, LINET can report several locations per flash in 3D (e.g. Stolzenburg et al., 2012), which requires that the closest sensor is within about 100 km of the lightning discharge, and hence can operate as a crude imaging system. Information about other systems (including optical) can be found in Rakov and Uman (2003, Ch. 17), Cummins and Murphy (2009), Betz et al. (2009), Nag et al. (2015) and references therein. There are more than 60 lightning locating networks worldwide that operate in the VLF/LF range.

Besides the general characterization of each system, the available information on its performance characteristics is given with emphasis on those based on formal ground-truth studies published in the peer-reviewed literature.

8.4.1 Lightning Mapping Array (LMA), 60–66 MHz

LMA networks typically consist of 10–15 stations separated by 15–20 km and connected by wireless communication links to a central location (Thomas et al., 2004). Each station receives the lightning signals (from both cloud and CG flashes) in a locally unused television channel (usually TV channel 3, 60–66 MHz). Typical time resolution (measurement time window) is 80–100 μs, which is sufficient for mapping relatively slow leader processes. A larger time window, typically 400 μs, is used for real-time processing and display.

The location accuracy of the New Mexico LMA has been investigated experimentally using a sounding balloon carrying a VHF transmitter, airplane tracks, and observations of distant storms (Thomas et al., 2004). Simple geometric models for estimating the location uncertainty of sources both over and outside the network have also been developed. The model results were found to be a good estimator of the observed errors. Sources over the network at altitudes ranging from 6 to 12 km (well above the lower cloud boundary) were located with an uncertainty of 6–12 m rms in the horizontal and 20–30 m rms in the vertical, resulting in less than a 100 m 3D error for most located sources. Outside the network the location uncertainties increase with distance.

8.4.2 US National Lightning Detection Network (NLDN), 400 Hz–400 kHz

The NLDN consists of more than 100 stations separated by typically 300–350 km and mostly covering the contiguous USA. A combination of TOA and MDF locating techniques is employed. Both cloud and CG lightning discharges are reported. Classification is accomplished by applying field waveform criteria to individual magnetic field pulses. Generally, pulses wider than a certain threshold are interpreted as being produced by return strokes, while narrower pulses are attributed to cloud discharge activity. All positive events with estimated peak currents <15 kA are classified by the NLDN as cloud pulses. Peak currents are estimated from measured fields using an empirical formula based on rocket-triggered-lightning data, with the field peaks being adjusted to account for expected propagation effects (stronger than the inverse proportionality distance dependence). Further information on the evolution of the NLDN, its enabling methodology, and applications of NLDN data can be found in Rakov and Uman (2003, Ch. 17), Rakov (2005), Orville (2008), Cummins and Murphy (2009), Nag et al. (2014) and references therein.

CG stroke and flash detection efficiencies have been investigated, using video cameras, in Southern Arizona, Oklahoma, and Texas (Biagi et al., 2007). The stroke detection efficiency in Southern Arizona was estimated to be 76 percent (N = 3620), and in Texas/Oklahoma, it was 85 percent (N = 885). The corresponding flash detection efficiencies were 93 percent (N = 1097) and 92 percent (N = 367). Additionally, classification of lightning events as cloud or CG discharges was examined in this study, as well as in a similar study (but additionally using independent (LASA) electric field waveform measurements) in the Colorado–Kansas–Nebraska region (Fleenor et al., 2009).

CG stroke and flash detection efficiencies have been also investigated, using as the ground-truth rocket-triggered-lightning data, in the Florida region (Jerauld et al., 2005; Nag et al., 2011; Mallick et al., 2014a). From the latest (2004–12) study, the CG stroke and flash detection efficiencies were found to be 75 percent and 94 percent, respectively. Strokes in rocket-triggered flashes are similar to regular subsequent strokes (following previously formed channels) in natural lightning and, hence, the 75 percent stroke detection efficiency value is applicable only to regular negative subsequent strokes in natural lightning. The flash detection efficiency is expected to be an underestimate of the true value for natural negative lightning flashes, since first strokes typically have larger peak currents than subsequent ones. Zhu et al. (2016), using high-speed video and electric field data on 366 natural CG strokes in Florida, found the NLDN stroke detection efficiency to be 93 percent. The classification accuracy in their study was 91 percent.

Nag and Rakov (2012) examined electric field waveforms produced by 45 positive flashes containing 53 strokes. Out of these 53 strokes, the NLDN located 51 (96 percent), of which 48 (91 percent) were correctly identified and three return strokes were misclassified as cloud discharges.

According to Cummins and Murphy (2009), the NLDN cloud-flash detection efficiency (a flash was considered detected if at least one VLF/LF pulse produced by that flash was detected) was in the range of 10–20 percent, depending on local differences in distances between stations. From a more recent study based on using data from two VHF lightning imaging systems (LMAs) as a reference, Murphy and Nag (2015) reported the cloud-flash detection efficiency to be in the 50–60 percent range. Wilson et al. (2013) stated that the NLDN typically reports one to three cloud pulses per flash. Nag et al. (2010) examined wideband electric fields, electric and magnetic field derivatives, and narrowband VHF (36 MHz) radiation bursts produced by 157 compact intracloud discharges (CIDs; see Appendix 4). The NLDN located 150 (96 percent) of those CIDs and correctly identified 149 (95 percent) of them as cloud discharges. Zhu et al. (2016), using high-speed video and electric field data on cloud discharge activity in both cloud and cloud-to-ground flashes in Florida, found the NLDN detection efficiency for those events (sequences of cloud pulses) to be 37 percent (N = 95). The classification accuracy in their study was 91 percent (N = 35). Out of ten complete cloud flashes, five were detected and all of those five were correctly classified.

Mallick et al. (2014a) estimated, from comparison of NLDN-reported locations with the precisely known locations of triggered-lightning ground attachment points, the median location error to be 334 m, with the largest error being 8 km. Data acquired in 2004–12 were used.

Peak current estimation errors have been estimated from comparison of NLDN-reported peak currents with directly measured currents at the triggered-lightning channel base. In 2004–12, the median absolute value of current estimation error was 14 percent (Mallick et al., 2014a). The current estimation errors never exceeded 127 percent in absolute value. These results (also the location error results based on triggered-lightning data) are applicable only to regular negative subsequent strokes in natural lightning.

8.4.3 LIghtning detection NETwork (LINET), 1–200 kHz

The basic location method used in this system is TOA, although the magnetic field sensors provide arrival-angle information that is employed as a "plausibility check" on computed locations. Height information derived from the arrival time at the nearest reporting sensor is employed to assist in classification of processes in cloud flashes and in-cloud processes (e.g. preliminary breakdown) in CG flashes on the one hand and CG strokes on the other (near-ground locations are assumed to be associated with CG strokes and elevated ones with all the other processes). It is stated that the reliable separation of return strokes and cloud pulses can be achieved as long as the closest sensor is within about 100 km of the lightning discharge, which requires baselines of 200–250 km or less. Emphasis is placed on detection of low-amplitude signals of both cloud and CG lightning and recognition of thunderstorm cells for nowcasting purposes. Peak currents for processes in cloud flashes, in-cloud processes in CG flashes, and CG strokes are estimated assuming direct proportionality between the peak current and peak magnetic (or electric) field and inverse distance dependence of field peak. More information about LINET can be found in Betz et al. (2009) and references therein.

Similar to VHF channel imaging systems, it is not clear how to define the detection efficiency for LINET, which, in a sense, also maps evolution of lightning channels, although with a considerably smaller number of located sources per flash. Additionally, in-cloud processes (e.g. preliminary breakdown) in CG flashes are assigned to the cloud lightning category, which is apparently inconsistent with the traditional definitions of cloud flash as a lightning discharge without CG strokes and CG flash as a lightning discharge that consists of both in-cloud processes and CG strokes. This is probably immaterial for a number of applications, such as cell tracking and detection of severe weather.

The random location error is claimed to be approximately 150 m, but the existence of systematic errors is acknowledged. Betz et al. (2009) showed an example of 58 located strokes apparently terminated on an instrumented tower with an average location error of less than about 100 m, after compensating systematic errors that caused a location bias of about 200 m. Peak current estimation errors for LINET are unknown (no comparison with ground-truth data has been made to date).

8.4.4 US Precision Lightning Network (USPLN), 1.5–400 kHz

This network employs the VLF/LF TOA technique and consists of 100 electric field sensors covering the continental USA and other parts of North America. No formal performance testing studies regarding this system have been reported, but the operators of the system claim, apparently from the network simulation analysis, 95 percent stroke detection efficiency and 250 m typical location error throughout most of North America (>80 percent detection efficiency and <1 km location error in "key deployment areas" elsewhere in the world). Differentiation between cloud and CG processes is apparently accomplished by examining the frequency content and amplitude of the received signals. The field-to-current conversion procedure has not been formally described, nor is any information about testing its validity available.

8.4.5 Earth Networks Total Lightning Network (ENTLN), 1 Hz–12 MHz

The ENTLN sensors operate in a frequency range from 1 Hz to 12 MHz (spanning the ELF, VLF, LF, MF, and HF ranges, where ELF indicates the extremely low-frequency range, from 3 to 30 Hz). The TOA method is employed. According to Heckman and Liu (2010), the whole electric field waveforms are used in both locating the lightning events and differentiating between cloud and CG processes. Strokes (or individual cloud events) are clustered into a flash if they are within 700 ms and 10 km of the first detected stroke (or cloud event). A flash that contains at least one return stroke is classified as a CG flash; otherwise, it is classified as a cloud flash. In the cell tracking and thunderstorm alert generation, only flashes (which are less likely than strokes to be missed by the system) are used. The system operates in the USA and in a number of other countries.

The operators of the system claim 40–50 percent cloud flash detection efficiency across much of the USA and up to 95 percent in the US Midwest and East (Heckman and Liu,

2010). Maximizing the detection efficiency for cloud flashes appears to be the primary focus of this system. Peak currents for processes in cloud flashes, in-cloud processes (e.g. preliminary breakdown) in CG flashes, and CG strokes are estimated assuming direct proportionality between the peak current and peak electric field and inverse distance dependence of field peak.

Mallick et al. (2015) have evaluated the performance characteristics of the ENTLN using, as ground truth, data for 245 negative return strokes in 57 flashes triggered from June 2009 to August 2012 at Camp Blanding, Florida. The performance characteristics were determined both for the ENTLN processor that had been in service at the time of acquiring triggered-lightning data (June 2009 to August 2012) and for the new ENTLN processor, introduced in November 2012. So, evaluation for the new processor simulates ENTLN output as if the new processor were in service from June 2009 to August 2012. For the same ground-truth dataset and the same evaluation methodology, different performance characteristics for those two processors were obtained. For the old processor, flash detection efficiency was 77 percent, stroke detection efficiency was 49 percent, fraction of misclassified events was 61 percent, median location error was 631 m, and median absolute current estimation error was 51 percent. For the new processor, flash detection efficiency was 89 percent, stroke detection efficiency was 67 percent, fraction of misclassified events was 54 percent, median location error was 760 m, and median absolute current estimation error was 19 percent.

8.4.6 World Wide Lightning Location Network (WWLLN), 6–18 kHz

The WWLLN utilizes the time-of-group-arrival (TOGA) method (see Section 8.2.5) to locate lightning events. As of March 2012, WWLLN employed 57 sensors located on all continents, although, according to Dowden et al. (2002), global coverage could be in principle provided by as few as 10 sensors. Distances between the sensors are up to thousands of kilometers. Presently, only those lightning events that triggered at least five sensors and that had residuals (uncertainties in the stroke timing) less than or equal to 30 µs are regarded as located with acceptable accuracy.

In their study of WWLLN performance characteristics, Abarca et al. (2010) used NLDN data as the ground truth and found that the CG flash detection efficiency increased from about 3.88 percent in 2006–7 to 10.3 percent in 2008–9, as the number of sensors increased from 28 in 2006 to 38 in 2009. For events with NLDN-reported peak currents of 130 kA or higher, the detection efficiency was reported to be 35 percent. The average location error was estimated to be 4–5 km. Hutchins et al. (2012a) developed a model to compensate for the uneven global coverage of the WWLLN. It is known that field peaks at distances >700 km or so are due to ionospheric reflections rather than the ground wave. Interaction of lightning signals with the ionosphere spectrally distorts the field waveform so that it is not straightforward to infer the peak current and even polarity of lightning. Nevertheless, Hutchins et al. (2012b) developed a method to convert the stroke-radiated power in the 6–18 kHz band to peak current.

Mallick et al. (2014) evaluated the performance characteristics of the WWLLN using rocket-triggered lightning data acquired at Camp Blanding, Florida, in 2008–13. The flash and stroke detection efficiencies were 8.8 percent and 2.5 percent, respectively. The stroke detection efficiency for strokes with peak current ≥ 25 kA was 29 percent. The median location error was 2.1 km. The median absolute error in WWLLN peak currents estimated from the empirical formula of Hutchins et al. (2012b) was 30 percent.

8.4.7 Global Lightning Dataset (GLD360), 300 Hz–48 kHz

The Global Lightning Dataset (GLD360), also referred to as the Global Lightning Detection Network (GLDN), employs an unspecified number of VLF sensors strategically placed around the world. As stated in Section 8.2.5, locations are obtained using both TOA and MDF methods in conjunction with a lightning waveform recognition algorithm. A lightning event must be detected by at least three sensors to be located. The system does not distinguish between ground and cloud lightning events.

Demetriades et al. (2010) evaluated the GLD360 performance characteristics using NLDN data as the ground truth and found that the CG flash detection efficiency was 86–92 percent, and the median location error was 10.8 km. From a similar study, but using the Brazilian lightning detection network, Naccarato et al. (2010) reported the CG flash detection efficiency of 16 percent and the mean location error of 12.5 km. GLD360 performance in Europe in May–September 2011 was compared with that of the networks participating in the European Cooperation for Lightning Detection (EUCLID) by Ponjola and Makela (2013). Poelman et al. (2013), using electric field measurements in conjunction with high-speed video recordings for 210 strokes in 57 negative CG flashes in Belgium, estimated the flash and stroke detection efficiencies to be 96 and 70 percent, respectively. They also estimated the median location error of 1.3 km (N = 134) relative to EUCLID locations. Said et al. (2013), using NLDN data as the ground truth, estimated the ground flash detection efficiency of 57 percent and median location error of 2.5 km.

The GLD360 also reports the peak current (inferred from the measured magnetic field peak) and polarity. The latter is determined via the cross-correlation with the bank of "canonical" waveforms. Said et al. (2013) estimated the mean and geometric mean errors in peak current estimates relative to NLDN-reported peak currents to be 21 and 6 percent, respectively. They also found that for 96 percent of matched events, GLD360 reported the same polarity as the NLDN.

Mallick et al. (2014b) used rocket-triggered lightning data acquired in 2011–13 at Camp Blanding, Florida, to evaluate the GLD360 performance characteristics. The flash and stroke detection efficiencies were 67 and 37 percent, respectively. Out of 75 detected strokes, one (1.3 percent) was reported with incorrect polarity. The median location error was 2.0 km, and the median absolute current estimation error was 27 percent.

8.5 Summary

There exists a variety of lightning locating techniques that are based on the detection of lightning radio-frequency electromagnetic signals, with accurate locating being possible only by using multiple-station systems. When a single location per cloud-to-ground light-ning stroke, typically the ground strike point, is required, magnetic direction finding, the long-baseline time-of-arrival technique, or a combination of the two can be employed. Location accuracies of the order of hundreds of meters and flash detection efficiencies of about 90 percent are possible. When electromagnetic imaging of the developing channels of any type of lightning flash is required, the VHF time-of-arrival technique or VHF inter-ferometry can be used. Lightning locating systems operating on global scale utilize methods that are capable of extracting source information from electromagnetic signals dominated by ionospheric reflections.

Although theoretical models can be used to evaluate the LLS performance character-istics, ultimately ground-truth data are needed to verify model-predicted results. Such data should include the measured time, position, and peak current of lightning discharges in a region covered by LLS and can be obtained using instrumented towers or rocket-triggered lightning. Video camera observations can be used for evaluating detection efficiency and classification of lightning events (CG vs. non-CG).

Questions and problems

8.1. Explain the 180° ambiguity in magnetic direction finding systems.
8.2. Why are the time-of-arrival systems sometimes referred to as hyperbolic?
8.3. What is the minimum number of stations needed to obtain unique location with (a) MDF technique and (b) TOA technique?
8.4. What is LMA?
8.5. Give examples of global lightning locating systems. In what frequency range do they operate?
8.6. List the performance characteristics of lightning locating systems.
8.7. Discuss the use of rocket-triggered lightning for testing the performance characteristic of lightning locating systems.
8.8. What is the detection efficiency? How can it be verified for (a) cloud-to-ground flashes and (b) cloud flashes?

Further reading

Betz, H. D., Schumann, U., and Laroche, P., eds. (2009). *Lightning: Principles, Instruments and Applications*, 691 pp., Springer.
Rakov, V. A. and Uman, M. A. (2003). *Lightning: Physics and Effects*, 687 pp., New York: Cambridge.

Lightning damaging effects and protective techniques

9.1 General principles

Systematic studies of thunderstorm electricity can be traced back to May 10, 1752 in the village of Marly-la-Ville, near Paris. On that day, in the presence of a nearby storm, a retired French dragoon, acting on instructions from Thomas-Francois Dalibard, drew sparks from a tall iron rod that was insulated from ground by wine bottles. The results of this experiment, proposed by Benjamin Franklin, provided the first direct proof that thunderclouds contain electricity. Even before the experiment at Marly, Franklin had proposed the use of grounded rods for lightning protection. Originally, he thought that the lightning rod would silently discharge a thundercloud and thereby would prevent the initiation of lightning. Later, Franklin stated that the lightning rod had a dual purpose: if it cannot prevent the occurrence of lightning, it offers a preferred attachment point for lightning and then a safe path for the lightning current to ground. It is in the latter manner that lightning rods, often referred to as Franklin rods, actually work.

There are generally two aspects of lightning protection design: (1) diversion and shielding, primarily intended for structural protection, but also serving to reduce the lightning electric and magnetic fields within the structure, and (2) the limiting of currents and voltages on electronic, power, and communication systems via surge protection. Primarily the first aspect will be considered in this chapter. Properly designed structural lightning protection systems for ground-based structures serve to provide lightning attachment points and paths for the lightning current to follow from the attachment points into the ground without harm to the protected structure. Such systems are basically composed of three elements: (1) "air terminals" at appropriate points on the structure to intercept the lightning, (2) "down conductors" to carry the lightning current from the air terminals toward the ground, and (3) "grounding electrodes" (ground terminals) to pass the lightning current into the earth. The three system components must be electrically well connected. The efficacy of this so-called conventional approach to lightning protection is evidenced by the comparative statistics of lightning damage to protected and unprotected structures. Neither data nor theory supports claims that non-conventional approaches, including "lightning elimination" and "early streamer emission" techniques, are superior to the conventional one (Uman and Rakov, 2002).

A lightning protection system for houses proposed in 1778 is shown in Fig. 9.1. For comparison, modern structural lightning protection is illustrated in Fig. 9.2. Clearly, the

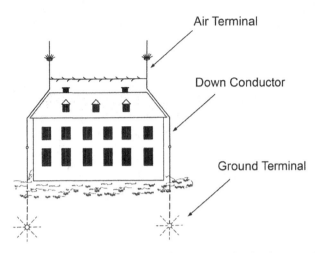

Fig. 9.1   Lightning protection system for houses proposed (most likely by G. Ch. Lichtenberg) in 1778. Adapted from Wiesinger and Zischank (1995).

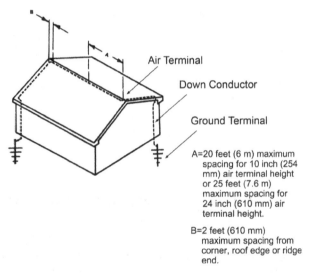

Fig. 9.2   Modern structural lightning protection (UL 96A, 1998).

approach remains essentially the same. Note that metallic roofs whose thickness is 4.8 mm (3/16 in) or greater do not require air terminals (NFPA 780).

In 1876, James Clerk Maxwell suggested that Franklin rod systems attracted more lightning strikes than the surrounding area (actually, those systems just intercept imminent strikes to the structure to be protected). He proposed that a gunpowder building be completely enclosed with metal of sufficient thickness, forming what is now referred to as a Faraday cage. If lightning were to strike a metal-enclosed building, the current would be constrained to the exterior of the

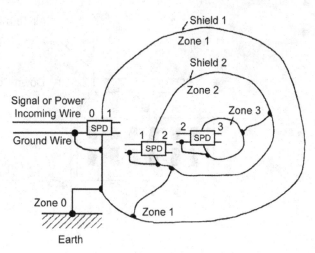

Fig. 9.3 The general principles of topological shielding. SPD stands for "surge protective device." Adapted from Vance (1980).

metal enclosure, and it would not even be necessary to ground this enclosure. In the latter case, the lightning would merely produce an electric arc from the enclosure to the earth. The Faraday cage effect is provided by all-metal cars and airplanes. Modern steel-frame buildings with reinforcing metal bars in the concrete foundation connected to the building structural steel provide a good approximation to a Faraday cage. As the spacing between conductors increases, however, the efficiency of the lightning protection decreases. In practice, a combination of the Franklin rod system concept and the Faraday cage concept is often used. Modern lightning protection schemes for structures containing computers or other sensitive electronics employ a technique known as topological shielding with surge suppression (see Fig. 9.3), which can be viewed as a generalization of the Faraday cage concept.

9.2 Lightning parameters vs. damage mechanisms

The type and amount of lightning damage that an object suffers depends on both the characteristics of the lightning discharge and the properties of the object. The physical characteristics of lightning of most interest are various properties of the current waveform and of the radio-frequency electromagnetic fields. Four distinct properties of the lightning current waveform are considered important in producing damage: (1) the peak current, (2) the maximum rate of change of current, (3) the integral of the current over time (i.e. the charge transferred), and (4) the integral of the current squared over time, the "action integral." Each of these properties and the type of damage to which it is thought to be related are discussed by Rakov and Uman (2003, Ch. 18; see also Chapter 4 of this book) and summarized below.

(1) Peak current. For objects or systems that present an essentially resistive impedance, such as, under certain conditions, a ground rod driven into the earth, a long power line, and a tree, the voltage (V) on the object or system with respect

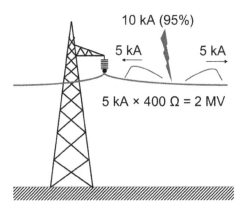

Fig. 9.4   Schematic representation of direct lightning strike to a power line phase conductor with 400 Ω characteristic impedance. A 10 kA peak current (95 percent of lightning first-stroke currents exceed this value, according to the IEEE distribution) is injected into the conductor. It produces a voltage of 2 MV between the phase conductor and the grounded metallic tower, which is likely to cause a flashover along the surface of the insulator string at the tower.

to remote ground will be proportional to the current, I, via Ohm's law, $V = RI$, where R is the effective resistance at the strike point. For example, a 10 kA peak current injected into a power line phase conductor with a 400 Ω characteristic impedance produces a voltage of 2 MV between the phase conductor and the grounded supporting tower, as illustrated in Fig. 9.4. Such a large voltage is likely to lead to an electric discharge (flashover) along the surface of the power trans- mission line insulator string, thereby providing a low-impedance path for the power (or industrial-frequency; usually 50 or 60 Hz) current to flow from the- struck phase conductor to ground. The resultant short-circuit fault can lead to an outage of the power line.

(2) Maximum rate-of-change of current (dI/dt). For objects that present an essentially inductive impedance, such as, under some circumstances, wires in an electronic system, the peak voltage will be proportional to the maximum rate-of-change (Usually rate-of-rise) of the lightning current ($V = L \, dI/dt$, where L is the inductance of the wire and V is the voltage difference between the two ends of the wire). For example, if a "ground" wire connecting two electronic systems (for example, in a communications tower and in an adjacent electronics building) has an inductance per unit length of 10^{-6} H m^{-1} and if 10 percent of the lightning current flows in the wire producing $dI/dt = 10^{10}$ A s^{-1}, 10 kV will appear across each meter of the wire. Thus, even a small fraction of the lightning current circulating in grounding and bonding wires can cause damage to solid-state electronic circuits that have communication, power, and other inputs "grounded" at different locations. Also, lightning induced effects are often related to dI/dt, as illustrated in Fig. 9.5.

(3) Integral of the current over time (charge transfers). The severity of heating or burn- through of metal sheets such as airplane wing surfaces and metal roofs is, to a first approximation, proportional to the lightning charge transferred, which is in turn

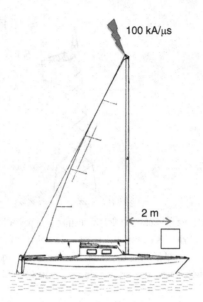

100 kA/µs

2 m

Fig. 9.5 Schematic representation of the lightning induced effect in a small rectangular loop of wire located 2 m from the metallic mast of a sailboat. The loop is vertical and its plane is oriented toward the mast. The mast is directly struck by lightning whose current rate of rise (dI/dt) is 100 kA/us. According to Faraday's Law, the magnitude of voltage induced in the loop (per unit area) is V/A = dB/dt, where A is the loop area, $B = \mu_0 I/(2\pi r)$ is the magnetic flux density (approximately constant over area A), and $\mu_0 = 4\pi \times 10^{-7}$ H/m is the permeability of free space. The resultant induced voltage V/A = 10 kV/m^2. If the loop area is 10 × 10 cm^2 (0.01 m^2), the voltage will be 100 V, enough to cause damage to sensitive electronic components.

proportional to the energy delivered to the surface. This is the case because the input power to the conductor surface is the product of the current and the more or less current-independent voltage drop at the arc-metal interface, this voltage drop being typically 5 to 10 V. Generally, large charge transfers are due to long duration (tens to hundreds of milliseconds) lightning currents, such as long continuing currents, whose magnitude is in the 100 to 1,000 A range, rather than return strokes having larger currents but relatively short duration and hence producing relatively small charge transfers. Additionally, even those impulsive current components, which do have relatively large charge transfers, cause only relatively minor surface damage on metal sheets, apparently because the current duration is too short to allow penetration of heat into the metal. Thermal damage caused by multiple lightning strikes to the metallic rod that was installed at the top of the 540 m high tower in Moscow, Russia, is shown in Fig. 9.6.

(4) Action integral. The heating and melting of resistive materials, which may or may not be relatively good conductors, and the explosion of poorly conducting materials are, to a first approximation, related to the value of the action integral; that is, the time integral of the Joule heating power $I^2(t) R$, for the case that R = 1 Ω. Thus, the action integral (also referred to as the specific energy) is a measure of the ability of the lightning current to generate heat in a strike object characterized by a resistance R. In the case of

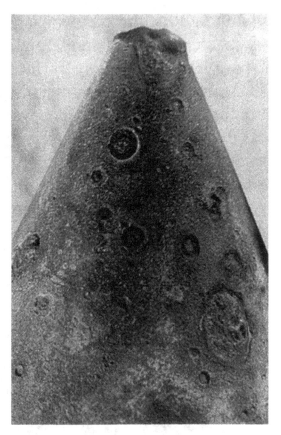

Fig. 9.6 Lightning damage to the steel tip of the 540 m high Ostankino Television Tower in Moscow, Russia. Adapted from Bazelyan and Raizer (2000).

the object having a relatively high resistance, such as a tree, this heat vaporizes the internal material and the resultant gas pressure causes an explosive effect. An example of lightning damage to a tree is shown in Fig. 9.7. In addition to heating effects, the action integral is also a measure of some mechanical effects such as the crushing of hollow metal tubes carrying lightning current, the effect being both a function of the instantaneous force, which is proportional to the square of the current and the duration of application of the force. In this case, the applied force must also exceed some threshold value.

Electromagnetic fields from lightning that impinge on any conducting objects induce currents and resultant voltages in those objects. Two properties of the electromagnetic fields are sufficient to describe most of the important damaging effects, commonly the destruction of electronic components: (1) the peak values of the electric and magnetic fields and (2) the maximum rates-of-change of the fields. For certain types of unintended antennas, such as elevated conductors that are capacitive coupled to ground, the peak-induced voltage on the conductors with respect to ground is proportional to the peak electric field. For other

Fig. 9.7   Damage to the oak tree struck by lightning on March 29, 2009 in Gainesville, Florida. Photograph by Christopher Hayes.

unintended antennas, such as a loop of wire in an electronic circuit, some underground communication cables, and elevated conductors resistively coupled to ground, the peak voltage is proportional to the maximum rate of change of the electric or the magnetic field. The degree of coupling of fields through holes or apertures in the metal skins of aircraft and spacecraft is generally proportional to the rate of change of the electric and magnetic fields.

9.3 Electrogeometrical model (EGM)

The attachment of the leader to the strike object is often described using the so-called electrogeometrical model (EGM), the core of which is the concept of a "striking distance."

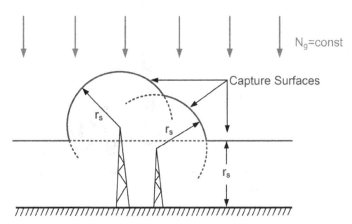

Fig. 9.8 Illustration of capture surfaces of two towers and earth's surface forming the overall capture surface in the electrogeometrical model (EGM). r_s is the striking distance. Vertical arrows represent descending leaders, assumed to be uniformly distributed (N_g = const) above the capture surfaces. Adapted from Bazelyan and Raizer (2000).

This concept obscures some of the significant physics but allows the development of relatively simple and useful techniques for designing lightning protection systems for various structures. The striking distance can be generally defined as the distance from the tip of the descending leader to the object to be struck at the instant when the lightning termination point is uniquely determined. There are different specific definitions of striking distance. The simplest of them is based on the assumption that the descending leader is at striking distance from the strike object when an upward connecting leader is initiated from this object. Another definition requires the establishment of streamer connection between the downward and upward connecting leader channels. For a given striking distance, one can define an imaginary surface above the ground and above objects on the ground (see Fig. 9.8) such that, when the descending leader passes through that surface at a specific location, the leader is "captured" by a specific point on the ground or on a grounded object. The geometrical construction of this capture surface can be accomplished simply by rolling an imaginary sphere of radius equal to the assumed striking distance across the ground and across objects on the ground, the so-called rolling sphere method (RSM) (e.g. Lee, 1978; NFPA 780). The locus of all points traversed by the center of the rolling sphere forms the imaginary capture surface referred to above. Those points that the rolling sphere touches can be struck, according to this approach; and points where the sphere does not touch cannot. Figure 9.9 illustrates the rolling sphere method. The shaded area in Fig. 9.9 is that area into which, it is postulated, lightning cannot enter.

In the rolling sphere method, the striking distance is assumed to be the same for any object projecting above the earth's surface and for the earth itself. There are variations of the EGM in which the assumption of different striking distances for objects of different geometry is used (e.g. Eriksson, 1987). The main application of the rolling sphere method is positioning air terminals on an ordinary structure, so that one of the terminals, rather than a roof edge or other part of the structure, initiates the upward leader that intercepts the

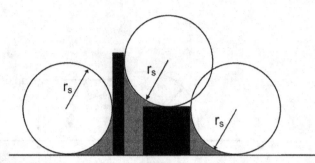

Fig. 9.9   Illustration of the rolling sphere method for two objects shown in black. r_s is the striking distance. The shaded area is that area into which, it is postulated, lightning cannot enter. Adapted from Szczerbinski (2000).

descending leader and, hence, becomes the lightning attachment point. According to the theory, for a given rolling sphere radius, all flashes with first-stroke peak current values higher than the corresponding minimum peak current value will be intercepted by the air terminals. For example, the NFPA 780 recommends the rolling-sphere radius of 45 m. The corresponding peak current is 10 kA, whose probability to be exceeded is 95 percent (according to the IEEE distribution). Thus, it is expected that the lightning protective system designed using the 45 m radius rolling sphere will protect against first strokes with peak currents exceeding 10 kA; that is, against 95 percent of all first strokes. For a more conservative design a smaller rolling-sphere radius should be used.

The striking distance is usually expressed as a function of prospective return-stroke peak current. The procedure to obtain such an expression basically involves assumptions of leader geometry, total leader charge, distribution of charge along the leader channel, and critical average electric field between the leader tip and the strike object at the time the connection between them becomes imminent. This critical electric field is assumed to be equal to the average breakdown field from long laboratory spark experiments with rod-rod and rod-plane gaps, which varies with waveshape of applied voltage as well as with other factors such as the high-voltage generator circuitry. The typical assumed values range from 200 to 600 kV/m. As a result, one can obtain an expression relating the striking distance to the total leader charge. In the next step, the observed relation (see Fig. 9.10) between the charge and resultant return-stroke peak current (Berger, 1972) is used to express the striking distance, r_s, in terms of the peak current, I. The most popular striking-distance expression, included in many lightning protection standards, is

$$r_s = 10 \times I^{0.65} \tag{9.1}$$

where I is in kA and r_s is in meters. This and other expressions for the striking distance found in the literature are illustrated in Fig. 9.11. Given all the assumptions involved and large scatter in the experimental charge-current relation (see Fig. 9.10), each of these relationships is necessarily crude, and the range of variation among the individual expressions (see Fig. 9.11) is up to a factor of 3 or more. Therefore, there are considerable uncertainties in estimating the striking distance. On the other hand, there is satisfactory long-term experience with the RSM as applied to placement of lightning rods on ordinary

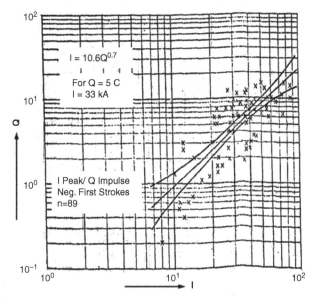

Fig. 9.10 Scatter plot of impulse charge, Q, versus return-stroke peak current, I, for first strokes in 89 negative flashes in Switzerland. Note that both vertical and horizontal scales are logarithmic. The best fit to data, $I = 10.6\ Q^{0.7}$, where Q is in coulombs and I is in kiloamperes, was used in deriving striking distance equations. Adapted from Berger (1972).

structures (the RSM has been in the Hungarian Standard on Lightning Protection since 1962; Horvath, 2000) and with the EGM in general as applied to power lines. This experience is the primary justification for the continuing use of this method in lightning protection studies.

The EGM can be used for estimating lightning incidence to different elements (usually to the protected object) of a structure as follows. One needs to (1) assume the spatial distribution of descending lightning leaders above all the capture surfaces (see Fig. 9.8) and specify the ground flash density, N_g (typically $N_g = \text{const}$), (2) find the striking distance, $r_s(I)$, and then the projection, $S(I)$, of the resultant capture surface of the element in question onto the ground surface, (3) specify the statistical distribution (the probability density function, to be exact) of lightning peak currents, $f(I)$, and (4) integrate the product $N_g \times S(I) \times f(I) \times dI$ from 0 to I_{max}, to obtain the lightning incidence (number of strikes per year) to the element in question. Alternatively, one can eliminate the finding of $S(I)$ in item (2) and entire item (4) from the outlined procedure using the Monte Carlo technique.

9.4 Bonding vs. isolating approaches in lightning protection practice

The twofold objective of structural lightning protection is (1) to force the current flow where one wants it to go and (2) not to allow the development of hazardous

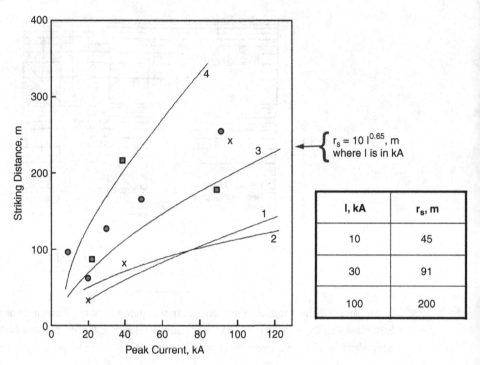

Fig. 9.11 Striking distance versus return-stroke peak current [curve 1, Golde (1945); curve 2, Wagner (1963); curve 3, Love (1973); curve 4, Ruhling (1972); x, theory of Davis (1962); ⦿, estimates from two-dimensional photography by Eriksson (1978); ▣, estimates from three-dimensional photography by Eriksson (1978)]. Adapted from Golde (1977) and Eriksson (1978).

potential differences. A difference of potential of 2 MV is sufficient for a side flash of over 1.8 m in air (NFPA 780) or arcing through the soil over 5.4 m or so. Once arcing takes place, an unplanned and uncontrolled current path is created. The arc is likely to turn moisture in the soil or structural material to steam with potentially damaging steam explosion effects. Destructive arcing between exposed or buried elements of the lightning protective system (LPS) and nearby metallic objects can be prevented by either (1) equipotential bonding or (2) adequate electrical isolation. If direct bonding is not acceptable, it should be done via a surge protective device (SPD) with suitable characteristics.

Ideally, when lightning current causes a properly protected system's potential to rise momentarily to as much as some megavolts, all points of bonded conductors "rise" together (neglecting traveling-wave effects that occur on electrically long conductors), and no hazardous potential differences are created. This scenario is somewhat similar to that of a bird sitting on a high-voltage wire unaware that it is at a time-varying potential whose amplitude exceeds 1 kV.

The choice between the bonding and isolating approaches usually depends on whether it is possible or not to separate LPS conductors from other conductors of the system by

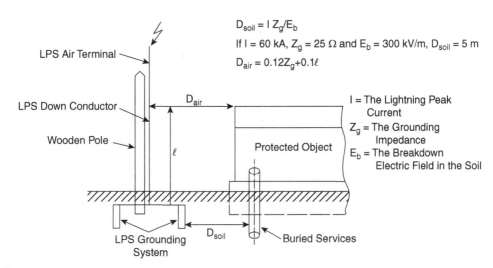

$$D_{soil} = I\, Z_g/E_b$$

If I = 60 kA, Z_g = 25 Ω and E_b = 300 kV/m, D_{soil} = 5 m

$$D_{air} = 0.12 Z_g + 0.1\ell$$

I = The Lightning Peak Current

Z_g = The Grounding Impedance

E_b = The Breakdown Electric Field in the Soil

LPS Air Terminal

LPS Down Conductor

Wooden Pole

Protected Object

D_{air}

ℓ

LPS Grounding System

D_{soil}

Buried Services

Fig. 9.12   Illustration of safety distances in air (D_{air}) and in the soil (D_{soil}). Adapted from Kuzhekin et al. (2003).

distances that are larger than the so-called safety distance. This safety distance depends on the breakdown electric field (which is different in air and in the soil), the magnitude of the lightning current (also the current rate-of-rise when the inductance of down-conductors is involved), and the impedance "seen" by the lightning at its attachment point (which depends on the soil resistivity, the geometry of the grounding electrode, and, in the case of side flash in air, on the inductance between the point of interest and the grounding system).

According to Kuzhekin et al. (2003), the distance between an LPS down-conductor and the protected object in air should be greater than D_{air} (see Fig. 9.12) given in meters by

$$D_{air} = 0.12 \times Z_g + 0.1 \times \ell \ \ (m) \tag{9.2}$$

where Z_g is the impedance, in ohms, of LPS grounding system under direct lightning strike conditions (transient impedance, which can be either smaller or larger than the dc grounding resistance, R) and ℓ is the distance, in meters, between the point of interest and the LPS grounding system. For a point in the immediate vicinity of ground surface, $\ell \approx 0$ and Eq. 9.2 reduces to

$$D_{air} = 0.12 \times Z_g \ \ (m) \tag{9.3}$$

If Z_g = 25 Ω, D_{air} = 0.12 × 25 = 3 m. The distance between the LPS grounding system and buried metallic services should be greater than D_{soil} given by

$$D_{soil} = I \times Z_g/E_b \ \ (m) \tag{9.4}$$

where I is the lightning peak current, and E_b is the breakdown electric field in the soil. Kuzhekin et al. (2003) assumed I = 60 kA (approximately 10 percent value,

Table 9.1 D_{soil} given by Eq. 9.4 as a function of lightning peak current								
Peak current, I, kA	5	10	20	40	60	80	100	200
Percentage exceeding tabulated value, $P_I \cdot 100\%$	99	95	76	34	15	7.8	4.5	0.8
D_{soil}, m ($Z_g = 25\ \Omega$, $E_b = 300$ kV/m)	0.42	0.83	1.7	3.3	5.0	6.7	8.3	17

which is recommended for lightning protection studies in Russia) and $E_b = 300$ kV/m, so that

$$D_{soil} = 0.2 \times Z_g \ \ (m) \tag{9.5}$$

If $Z_g = 25\ \Omega$, $D_{soil} = 0.2 \times 25 = 5$ m. Whatever the value of Z_g, Kuzhekin et al. (2003) recommend that $D_{air} \geq 5$ m and $D_{soil} \geq 3$ m ($D_{air} > D_{soil}$ due to the inductance of down conductors). It follows that buried electrical cables within 3 m of an LPS grounding system must be bonded to that system. An insulated single-wire cable can be bonded either via SPD or via an enclosing metallic pipe. Kuzhekin et al. (2003) state that the capacitance between the cable and the pipe is sufficiently large, making the capacitive impedance between them under direct lightning strike conditions very small. As a result, the wire and the pipe become effectively bonded (at the high frequencies characteristic of lightning current), this effect saving cable's insulation from electrical breakdown.

It follows from Eq. 9.4 that the lightning safety distance in the soil depends on (1) the magnitude of the lightning peak current, (2) the breakdown electric field in the soil, and (3) the impedance "seen" by the lightning at its attachment point.

We now consider the role of the lightning peak current. Clearly, D_{soil} increases with increasing I. Lightning peak currents for first strokes vary by a factor of 50, from about 5 kA to 250 kA, or more (see Chapter 4). The probability of occurrence of a given value of peak current rapidly increases with increasing I, up to 25 kA or so, and then slowly decreases. In designing lightning protective schemes, it is customary to consider, as the "worst case," moderately severe strokes that still have an appreciable probability of occurrence, so that the object is protected against the overwhelming majority of strokes (up to the "worst case"). Table 9.1 gives values of D_{soil} computed using Eq. 9.4 for different values of I (corresponding values of P_I, computed using Eq. 4.9, are indicated), assuming that $Z_g = 25\ \Omega$ and $E_b = 300$ kV/m. It seems to be reasonable to consider lightning strokes with peak currents up to 60 kA, which constitute about 85 percent of the population. In this case, $D_{soil} = 5.0$ m. If one would like to assure protection against strokes up to 80 or 100 kA, which constitute about 92.2 or 95.5 percent of the population, respectively, the corresponding values of D_{soil} will be 6.7 or 8.3 m.

9.5 Summary

Traditional lightning parameters needed in engineering applications include lightning peak current, maximum current derivative (dI/dt), average current rate of rise, current rise time, current duration, charge transfer, and action integral (specific energy), all derivable from direct current measurements. Distributions of these parameters presently adopted by most lightning protection standards are largely based on measurements by K. Berger and coworkers in Switzerland. The lightning protection technique introduced by Benjamin Franklin has proven its effectiveness as evidenced by the comparative statistics of lightning damage to protected and unprotected structures. The rolling sphere method commonly used in the design of such systems is relatively crude, in part because of our insufficient understanding of the lightning attachment process, but it does represent a useful engineering tool for determining the number and positions of air terminals. Topological shielding with surge protection provides the optimal approach to both structural and surge protection, but it is often costly.

The choice between the bonding and isolating approaches usually depends on whether or not it is possible to separate LPS conductors and other conductors of the system by distances that are larger than the so-called safety distance. This safety distance depends on the breakdown electric field (which is different in air and in the soil), the magnitude of the lightning current (also the current rate-of-rise when the inductance of down-conductors is involved), and the impedance "seen" by the lightning at its attachment point (which depends on the soil resistivity, the geometry of the grounding electrode, and, in the case of side flash in air, on the inductance between the point of interest and the grounding system).

Questions and problems

9.1. Name the elements of the lightning protective system for ordinary structures and explain the function of each of them.

9.2. Describe the Faraday cage effect.

9.3. Relate the lightning damaging effects to their causative lightning parameters.

9.4. What is the lightning action integral, if the lightning energy dissipated in a 100 Ω resistance at the strike point is 10^7 J?

9.5. Define the striking distance and explain how it is related to the prospective return-stroke peak current.

9.6. What is the main difference between the rolling-sphere method and the "classical" electrogeometrical model?

9.7. Why are the metallic objects at distances smaller than the safety distance from LPS conductors required to be bonded to the LPS?

9.8. Suppose that the grounding impedance of LPS is 10 Ω and the lightning peak current is 50 kA. What are the safety distances (a) in air at a height of 1 m above ground and (b) in the soil?

Further reading

Cooray, V., ed. (2010). *Lightning Protection*, 1036 pp., London: IET.

Rakov, V. A. and Uman, M.A. (2003). *Lightning: Physics and Effects*, 687 pp., New York: Cambridge.

Uman, M. A. (2008.) *The Art and Science of Lightning Protection*, 240 pp., Cambridge University Press.

Appendix 1 How is lightning initiated in thunderclouds?

Maximum electric fields typically measured in thunderclouds (see Table 3.2 of Rakov and Uman (2003) and references therein) are 1–2×10^5 V/m (the highest measured value is 4×10^5 V/m), which is lower than the expected conventional breakdown field, of the order of 10^6 V/m. Two mechanisms of lightning initiation have been suggested. One relies on the emission of positive streamers from hydrometeors when the electric field exceeds 2.5–9.5×10^5 V/m, and the other involves high-energy cosmic ray particles and the runaway breakdown that occurs in a critical field, calculated to be about 10^5 V/m at an altitude of 6 km. Either of these two mechanisms permits, in principle, creation of an ionized region ("lightning seed") in the cloud that is capable of locally enhancing the electric field at its extremities. Such field enhancement is likely to be the main process leading to the formation (via conventional breakdown) of a hot, self-propagating lightning channel.

A1.1 Conventional breakdown

According to the conventional breakdown mechanism, lightning is initiated via the emission of positive corona from the surface of precipitation particles, highly deformed by strong electric fields in the case of raindrops, coupled with some mechanism whereby the electric field is locally enhanced to support the propagation of corona streamers. Positive streamers are much more likely to initiate lightning than negative ones because they can propagate in substantially lower fields. The most detailed hypothetical scenario of lightning initiation via conventional breakdown is described by Griffiths and Phelps (1976b), who consider a system of positive streamers developing from a point on a hydrometeor where the electric field exceeds the corona onset value of 2.5–9.5×10^5 V/m (2.5–9.5 kV/cm). The developing streamers are assumed to form a conical volume that grows longitudinally. The ambient electric field in the thundercloud required to support the propagation of corona streamers, E_0, was found by Griffiths and Phelps (1976a) from laboratory experiments to be 1.5×10^5 V/m (1.5 kV/cm) at about 6.5 km and 2.5×10^5 V/m (2.5 kV/cm) at about 3.5 km. If the ambient electric field is higher than E_0, the streamer system will intensify, carrying an increasing amount of positive charge on the propagating base of the cone, which simulates the positive streamer tips, and depositing an equally increasing amount of negative charge in the conical volume which represents the trails of the positive streamers. As a result, an asymmetric conical dipole is formed, which presumably can serve to enhance the existing electric field at the cone apex representing the origin of the positive streamers on the surface of the hydrometeor. Further, Griffiths and Phelps (1976b) suggest that several conical streamer systems may develop sequentially, each one passing into the debris of its predecessors, to achieve the field enhancement required for breakdown. For representative values of ambient electric field and E_0 at 6.5 km altitude, they report that a series of three to seven such systems can give rise to local enhancement of the ambient electric field up to 1.5×10^6 V/m (15 kV/cm) over a distance of a few meters, which is sufficient to ensure dielectric breakdown and, perhaps, eventually lead to the formation of the stepped leader. The number of passages required to achieve a certain field value depends on the assumed value of the potential of the streamer tip. Griffiths and Phelps (1976b) used a value of 10 kV, based on the results of laboratory experiments reported by Phelps (1974), for most of their calculations. These calculations involve an extrapolation from the relatively small (up to 1 m) gaps used in laboratory experiments to the relatively large distances (of the order of 100 m) over which streamers might travel in a thundercloud.

Loeb (1966) considered a parcel of air containing positively charged raindrops that is swept in the updraft toward the negative charge center to yield positive corona streamers from the raindrops. In this scenario, formation of positive streamers is facilitated by updraft reducing the separation between the oppositely charged regions in the cloud.

Nguyen and Michnowski (1996) considered the effects of many closely spaced hydrometeors in lightning initiation. Their hypothetical mechanism involves a bidirectional streamer development assisted by a chain of precipitation particles, as opposed to the scenario that invokes the propagation of positive streamers alone.

A1.2 Runaway breakdown

Gurevich et al. (1992, 1999) suggested that runaway electrons may play an important role in lightning initiation. In order to "run away," an electron must gain more energy from the electric field between collisions with air particles than it loses in a collision. The so-called break-even electric field, which must be exceeded for runaway to occur, depends on altitude. This field decreases exponentially with altitude due to exponential decrease in the air molecule density. At altitudes of 4–6 km, the break-even electric field is 1.0×10^5–1.5×10^5 V/m (1–1.5 kV/cm) (Gurevich et al. 2003), which is about an order of magnitude lower than the conventional breakdown field at these altitudes. The runaway breakdown requires the presence of initial electrons with energies exceeding 0.1–1 MeV. Such energetic electrons are produced in thunderclouds via collisions of very high energy (10^{15}–10^{16} eV or greater) cosmic ray particles with atmospheric nuclei. The flux of particles with energies $\geq 10^{16}$ eV is about 0.1 per km^2 per second (Eidelman et al., 2004, Ch. 24). So, for a thundercloud area of about 100 km^2, they may occur every 100 ms. The initial energetic electrons are sometimes referred to as cosmic ray secondaries. Each very high energy cosmic

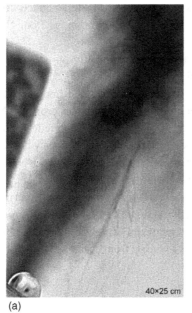 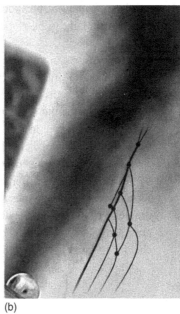

(a)   (b)

Fig. A1.1 (a) Infrared image of unusual plasma formation (UPF) recorded inside the artificial cloud of negatively charged water droplets. (b) Same as (a) but additionally showing a sketch of UPF superimposed on the image in order to improve UPF visualization. Note the cellular structure of the UPF channel network, which is not observed in leaders. Small solid circles in (b) schematically show positions of increased brightness, possibly corresponding to space stems facilitating the formation of the cellular network structure of UPF. Adapted from Kostinsky et al. (2015).

ray particle produces 10^6–10^7 secondaries (Gurevich and Zybin, 2001) in a process that is referred to as the "extensive air shower" (the term used in the cosmic ray physics for "cosmic ray shower").

If the energetic electrons find themselves in a thundercloud region in which the electric field is greater than the local break-even field (about 10^5 V/m at an altitude of 6 km), they can run away. Furthermore, if the high field region extends over a sufficient distance (of the order of a kilometer), an avalanche of runaway electrons and a very large number of relatively slow (a few eV to 100 keV or so) electrons can be produced. If the density of slow electrons reaches a critical value in some small region, the electrical conductivity in that region becomes high enough (of the order of 10^{-4} S/m; Solomon et al., 2001) to form an elongated plasma patch. This elongated (about 10 m long; Solomon et al., 2001) conductor, formed and polarized within microseconds (Gurevich and Zybin 2001), can enhance the electric field near its extremities to the values required for conventional breakdown via avalanches of low-energy (less than 30 eV) electrons. There has been a recent debate on whether a sufficient number of slow electrons can be produced to allow the creation of a plasma patch with conductivity of the order of 10^{-4} S/m (Dwyer and Babich, 2011, 2012; Gurevich et al., 2012). Note that the runaway breakdown initiated by an energetic cosmic ray particle (the initial energetic electron is supplied by an external source) is referred to as relativistic runaway breakdown, as opposed to the so-called cold runaway breakdown which is thought to be responsible for X-ray emissions from cloud-to-ground lightning leaders.

Recent efforts in studying the lightning initiation process are described in works of Gurevich and Karashtin (2013), Sadighi et al. (2015), and Kostinskiy et al. (2015). Specifically, Kostinskiy et al. (2015), using infrared imaging, observed unusual plasma formations inside the artificial cloud of charged water droplets. An example of one such formation is shown in Fig. A1.1. The authors reported that the unusual formations (1) occurred in charged cloud regions that did not previously host any detectable discharge activity and (2) contained hot channel segments in their overall network-like structure, while being distinctly different in their morphology from leaders. They speculated that such formations can be an intermediate stage between virgin air (in the presence of water droplets) or initial low conductivity streamer and a hot, self-propagating leader channel, provided that the observed hot segments can get polarized and grow within the overall channel network, thereby tapping energy from a relatively large cloud volume.

Appendix 2   Reconstruction of sources from measured electrostatic field changes

The measured electrostatic field changes for whole lightning flashes have been used to reconstruct both the overall charge sources for cloud and ground flashes and the separate charge sources for individual strokes and continuing current in ground flashes. In general, since the field is related to sources via an integral, the inverse problem of inferring sources from fields has an infinite number of solutions. It is possible to obtain a unique solution if a source model (type of charge distribution) is specified. The number of model parameters determines the minimum number of field measurement stations that are needed to obtain a solution. Three simple source models, monopole, dipole, and point dipole, are considered below.

Jacobson and Krider (1976) presented a least squares optimization method for fitting the parameters of assumed models of lightning-neutralized cloud charges to the electric field changes measured at multiple stations. That method has been generally adopted for the analysis of multiple-station electrostatic field changes measured using either field mill networks such as that at the Kennedy Space Center, Florida (e.g. Jacobson and Krider, 1976) or networks of ordinary electric field antennas with higher upper-frequency response (e.g. Krehbiel et al., 1979). The frequency response of a typical field mill is inadequate to resolve the field changes produced by individual lightning processes within a flash, only the total field change produced by the flash being measured. Source parameters are obtained by minimizing chi-squared (χ^2) function which is defined as

$$\chi^2 = \sum_{i=1}^{N} \frac{(\Delta E_{mi} - \Delta E_{ci})^2}{\sigma_i^2}, \tag{A2.1}$$

where ΔE_{mi} is the measured field change at the i-th ground station, ΔE_{ci} is the model-predicted (calculated) field change at the i-th station, σ_i^2 is the variance of the measurement at the i-th station due to experimental error, and N is the number of stations. The factor $1/\sigma_i^2$ can be viewed as a weighting factor for the data from the i-th station in the sum on the right-hand side of Eq. A2.1. Thus, that factor is a measure of the quality of the data from the i-th station.

A2.1 Monopole model

We will start with the simplest case in which the model-predicted field change ΔE_{ci} at each ground-based, $z_i = 0$, station is assumed to be due to the neutralization of a single,

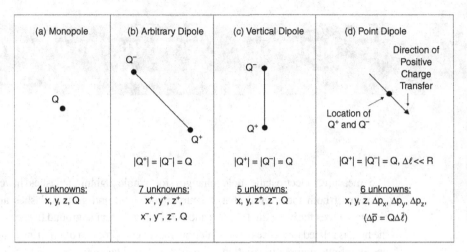

Summary of different models used to infer cloud charges neutralized by lightning discharges based on multiple-station measurements of ΔE. The monopole model (a) is applied to cloud-to-ground strokes, flashes, or continuing currents, while the three versions of the dipole model (b), (c), and (d) are used to represent cloud discharges, including the preliminary breakdown in CG flashes. In the point-dipole model, $\Delta\ell$ is the distance (unresolved in the model) between Q^+ and Q^-, which should be much smaller than the inclined distance R to either of them. Drawing by Manh D. Tran.

spherically-symmetric charge region whose unknown charge is Q and whose center is at an unknown location (x,y,z) above an assumed flat and perfectly conducting ground. Such a model is called the "monopole" or "point-charge" model (Fig. A2.1a) and the corresponding equation for ΔE_{ci} is

$$\Delta E_{ci} = \frac{2Qz}{4\pi\varepsilon_o[(x - x_i)^2 + (y - y_i)^2 + z^2]^{3/2}}, \qquad (A2.2)$$

which is a generalized form of Eq. 3.2. This equation can be applied to cloud-to-ground strokes, continuing currents (or portions of continuing currents), and ground flashes as a whole. There are four unknowns, x, y, z, and Q, in Eq. A2.2, and hence measurements at four or more stations are required. Additional measurements allow the evaluation of errors in estimating the unknowns. Equation A2.2 is substituted in Eq. A2.1 and the four unknowns x, y, z, and Q are iteratively adjusted until χ^2 is a minimum using, for example, the Marquardt algorithm described by Bevington (1969). The values of x, y, z, and Q corresponding to the χ^2 minimum are considered as the best fit to the measurements and the value of the χ^2 minimum as a measure of the adequacy of the fit. Jacobson and Krider (1976) multiplied the sum on the right-hand side of Eq. A2.1 by $1/\nu$ where ν is the number of degrees of freedom found as the number of measurements minus the number of unknowns in the model, the result being a normalized χ^2. Jacobson and Krider (1976) considered a solution valid when values of the normalized χ^2 at the χ^2 minimum were equal or less than 10.0.

A2.2 Dipole model

Eq. A2.2 can be easily extended (using the principle of superposition) to model cloud flashes or in-cloud processes of ground flashes by postulating that the calculated field change is due to the neutralization of two charges of equal magnitude but of opposite polarity: a dipole model. The number of unknowns in this case (arbitrarily oriented dipole; Fig. A2.1b) is seven, including three coordinates for each of the two charges and their magnitude. Accordingly, the minimum number of stations is seven. If the two charges are vertically stacked (see Section 3.2.4 and Fig. A2.1c), the x- and y-coordinates for both charges are the same and the minimum number of stations reduces to five.

A2.3 Point dipole model

When the spacing between the two charges of a dipole is small compared to the distance from the charges to each of the measurement stations, the so-called "point dipole" approximation (e.g. Krehbiel et al., 1979) can be used, in which

$$\Delta E_{ci} = \frac{1}{4\pi\varepsilon_0} \left[\frac{2\Delta p_z}{R_i^3} - \frac{6z}{R_i^5} \overline{R}_i \cdot \Delta\overline{p} \right], \tag{A2.3}$$

where $\Delta\overline{p} = Q\Delta\overline{l} = \Delta p_x \overline{a}_x + \Delta p_y \overline{a}_y + \Delta p_z \overline{a}_z$ is the vector dipole moment change with $\Delta\overline{l}$ being the vector distance from negative to positive charge, and $\overline{R}_i = (x - x_i)\overline{a}_x + (y - y_i)\overline{a}_y + z\overline{a}_z$ is the inclined distance vector from the source (point dipole) to the observer. There are six unknowns, $x, y, z, \Delta p_x, \Delta p_y, \Delta p_z$, and hence at least six measurements are required. Clearly, the point-dipole model provides less information than the dipole model.

The point-dipole solutions are usually presented graphically (see Figs. 3.4 and A2.1d) as arrows whose direction indicates the direction of the effective positive charge transport and whose length indicates the magnitude of the point-dipole moment change $|\Delta\overline{p}|$. The middle point on the arrow, usually marked by a dot in the graphical representation, indicates the single position for the neutralized positive and negative charges, with the actual positions being unresolved in this approximation.

In this appendix we will derive the exact electric and magnetic field equations for the case of the vertical lightning channel and perfectly conducting ground. We will start with the expressions of fields in terms of potentials:

$$\overline{E} = -\nabla\phi - \partial\overline{A}/\partial t$$

$$\overline{B} = \nabla \times \overline{A}$$

where $\overline{E}$ is the electric field intensity, $\overline{B}$ is the magnetic flux density, ϕ is the electric scalar potential, and $\overline{A}$ is the magnetic vector potential. The derivation will be performed in four steps corresponding to four configurations illustrated in Fig. A3.1: (a) differential current element at $z' = 0$ in free space; field point at $z > 0$, (b) elevated differential current element and its image; field point at $z > 0$, (c) same as (b), but the field point is at $z = 0$, (d) vertical lightning channel above ground; field point on the ground surface.

A3.1　Differential current element at $z' = 0$ in free space; field point at $z > 0$

Referring to Fig. A3.1a, the magnetic vector potential at point P due to a differential current element (Hertzian dipole) $I dz'$ pointing in the positive z-direction and located at $z' = 0$ in free space is given by

$$d\overline{A}(t) = \frac{\mu_0}{4\pi} \frac{I(0,\, t - R/c)}{R} dz'\, \overline{a}_z \tag{A3.1}$$

where $\overline{a}_z$ is the z-directed unit vector, and R/c is the time required for electromagnetic signal to propagate from the source to the field point at the speed of light, c. Note that for a cylindrical source of length dz' and volume dv', $I dz'$ is the same as $J dv'$, where J is the current density. The vector potential has the same direction as its causative current element. This is why $d\overline{A}(t)$ has only the z-component. In the following, in order to simplify notation, we will use $\overline{A}$ instead of $d\overline{A}(t)$ and drop 0 (indicating that $z' = 0$) in the argument of the current function.

The magnetic flux density due to a z-directed current element has only the ϕ-component (ϕ here is the azimuth, not to be confused with the scalar potential):

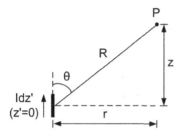

(a) Differential Current Element at z'=0
in Free Space; Field Point at z>0.

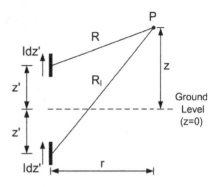

(b) Elevated Differential Current
Element (z'>0) and Its Image.

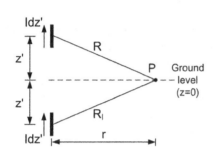

(c) Same as (b), but the Field Point is
at z=0.

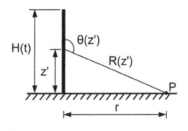

(d) Vertical Lightning Channel of Length
H(t) Above Ground; Field Point on the
Ground Surface (z=0).

Fig. A3.1   Four configurations used in deriving exact lightning electric and magnetic field equations. Drawing by Potao Sun.

$$d\overline{B} = \nabla \times \overline{A} = -\frac{\partial A_z}{\partial r}\overline{a}_\phi = -\frac{\mu dz'}{4\pi}\frac{\partial}{\partial r}\left[\frac{I(t - R/c)}{R}\right]\overline{a}_\phi \qquad (A3.2)$$

Since $R = (r^2+z^2)^{1/2}$, we use the product rule to take the derivative with respect to r. Then, noting that the derivative of 1/R with respect to r is $-r/R^3$, we get

$$d\overline{B} = -\frac{\mu dz'}{4\pi}\left[\frac{1}{R}\frac{\partial I(t - R/c)}{\partial r} + I(t - R/c)\left(-\frac{r}{R^3}\right)\right]\overline{a}_\phi \qquad (A3.3)$$

The spatial derivative $\partial I/\partial r$ can be converted to the time derivative $\partial I/\partial t$ by comparing the following two equations, in which the chain rule was used to take the partial derivatives with respect to r and t and I' stands for the derivative of I with respect to (t – R/c):

$$\frac{\partial I(t - R/c)}{\partial r} = \frac{\partial I(t - R/c)}{\partial(t - R/c)} \cdot \frac{\partial(t - R/c)}{\partial r} = -\frac{r}{cR}I'(t - R/c)$$

$$\frac{\partial I(t - R/c)}{\partial t} = \frac{\partial I(t - R/c)}{\partial(t - R/c)} \cdot \frac{\partial(t - R/c)}{\partial t} = I'(t - R/c)$$

Thus,

$$\frac{\partial I}{\partial r} = -\frac{r}{cR} \frac{\partial I}{\partial t}$$

and

$$d\overline{B} = \frac{\mu_0 dz'}{4\pi} \left[\frac{r}{cR^2} \frac{\partial I(t - R/c)}{\partial t} + \frac{r}{R^3} I(t - R/c) \right] \overline{a}_\phi \tag{A3.4}$$

Noting that $r/R = \sin\theta$, we can also write

$$d\overline{B} = \frac{\mu_0 dz'}{4\pi} \sin\theta \left[\frac{1}{cR} \frac{\partial I(t - R/c)}{\partial t} + \frac{I(t - R/c)}{R^2} \right] \overline{a}_\phi \tag{A3.5}$$

The first term, containing $\partial I/\partial t$, is the magnetic radiation field component and the second term, containing I, is the magnetostatic field component.

Next we derive, using the dipole technique (see Chapter 5), the electric field equation for the same differential current element at $z' = 0$. The total electric field is the negative of the sum of the gradient of ϕ and the time derivative of $\overline{A}$, with ϕ and $\overline{A}$ in this technique being related by the Lorentz condition (see Eq. 5.5):

$$\phi = -c^2 \int_{-\infty}^{t} \nabla \cdot \overline{A} d\tau \tag{A3.6}$$

Since the gradient operator (differentiation with respect to spatial coordinates) and integration over time are independent, we can write

$$\nabla\phi = -c^2 \int_{-\infty}^{t} \nabla(\nabla \cdot \overline{A}) d\tau \tag{A3.7}$$

In the following, we will obtain equations for $\nabla \cdot \overline{A}$, $\nabla(\nabla \cdot \overline{A})$, and $\partial\overline{A}/\partial t$.

Recall that $\overline{A}$ has only the z-component ($A_z \overline{a}_z$) and, hence, the divergence of $\overline{A}$ has only one term:

$$\nabla \cdot \overline{A} = \frac{\partial A_z}{\partial z} = \frac{\mu_0 dz'}{4\pi} \frac{\partial}{\partial z} \left[\frac{I(t - R/c)}{R} \right] = \frac{\mu_0 dz'}{4\pi} \left[\frac{1}{R} \frac{\partial I(t - R/c)}{\partial z} - \frac{z}{R^3} I(t - R/c) \right] \tag{A3.8}$$

where $-z/R^3$ is the derivative of $1/R$ with respect to z.

Due to the cylindrical symmetry of the problem, $\nabla \cdot \overline{A}$ is independent of azimuth and $\nabla(\nabla \cdot \overline{A})$ has only radial (r) and vertical (z) components, so that the gradient operator can be written as

$$\nabla = \overline{a}_r \frac{\partial}{\partial r} + \overline{a}_z \frac{\partial}{\partial z}$$

Thus,

$$\nabla\phi = -c^2 \int_{-\infty}^{t} \left[\overline{a}_r \frac{\partial(\nabla\cdot\overline{A})}{\partial r}\right] d\tau - c^2 \int_{-\infty}^{t} \left[\overline{a}_z \frac{\partial(\nabla\cdot\overline{A})}{\partial z}\right] d\tau = \nabla\phi_r + \nabla\phi_z \qquad (A3.9)$$

In the following, we expand the integrands in Eq. A3.9, convert the spatial derivatives to time derivatives, expand second-order derivatives, and then assemble final expressions for $\nabla\phi_r$ and $\nabla\phi_z$ for the differential current element at $z' = 0$.

Expand the integrands:

$$\overline{a}_r \frac{\partial}{\partial r}(\nabla\cdot\overline{A}) = \frac{\mu_0 dz'}{4\pi} \frac{\partial}{\partial r}\left[\frac{1}{R}\frac{\partial I(t-R/c)}{\partial z} - \frac{z}{R^3}I(t-R/c)\right]\overline{a}_r \qquad \text{\underline{r-component}}$$

$$\frac{\partial}{\partial r}\left[\frac{1}{R}\frac{\partial I(t-R/c)}{\partial z}\right] = \frac{1}{R}\frac{\partial^2 I(t-R/c)}{\partial r\partial z} + \left(-\frac{r}{R^3}\right)\frac{\partial I(t-R/c)}{\partial z} \qquad \text{1st term}$$

$$\frac{\partial}{\partial r}\left[-\frac{z}{R^3}I(t-R/c)\right] = -\frac{z}{R^3}\frac{\partial I(t-R/c)}{\partial r} - z\left(-\frac{3r}{R^5}\right)I(t-R/c) \qquad \text{2nd term}$$

$$\overline{a}_z \frac{\partial}{\partial z}(\nabla\cdot\overline{A}) = \frac{\mu_0 dz'}{4\pi} \frac{\partial}{\partial z}\left[\frac{1}{R}\frac{\partial I(t-R/c)}{\partial z} - \frac{z}{R^3}I(t-R/c)\right]\overline{a}_z \qquad \text{\underline{z-component}}$$

$$\frac{\partial}{\partial z}\left[\frac{1}{R}\frac{\partial I(t-R/c)}{\partial z}\right] = \frac{1}{R}\frac{\partial^2 I(t-R/c)}{\partial z^2} + \left(-\frac{z}{R^3}\right)\frac{\partial I(t-R/c)}{\partial z} \qquad \text{1st term}$$

$$\frac{\partial}{\partial z}\left[-\frac{z}{R^3}I(t-R/c)\right] = \frac{\partial(-z)}{\partial z}\frac{1}{R^3}I(t-R/c) + \frac{\partial}{\partial z}\left(\frac{1}{R^3}\right)(-z)I(t-R/c) \qquad \text{2nd term}$$

$$+ \frac{\partial I(t-R/c)}{\partial z}\left(-\frac{z}{R^3}\right)$$

$$= -\frac{1}{R^3}I(t-R/c)(-z)\left(-\frac{3z}{R^5}\right)I(t-R/c) - \frac{z}{R^3}\frac{\partial I(t-R/c)}{\partial z}$$

Convert the spatial derivatives to time derivatives:

$$\frac{\partial}{\partial r}[I(t-R/c)] = -\frac{r}{cR}\frac{\partial}{\partial t}[I(t-R/c)]$$

$$\frac{\partial}{\partial z}[I(t-R/c)] = -\frac{z}{cR}\frac{\partial}{\partial t}[I(t-R/c)]$$

Expand the second order derivatives:

$$\frac{\partial^2 I(t-R/c)}{\partial z^2} = \frac{\partial}{\partial z}\left[-\frac{z}{cR}\frac{\partial I(t-R/c)}{\partial t}\right]$$

$$= \frac{\partial(-z)}{\partial z}\frac{1}{cR}\frac{\partial I(t-R/c)}{\partial t} + \frac{\partial}{\partial z}\left(\frac{1}{cR}\right)(-z)\frac{\partial I(t-R/c)}{\partial t} + \frac{\partial^2 I(t-R/c)}{\partial z\partial t}(-z)\frac{1}{cR}$$

$$= -\frac{1}{cR}\frac{\partial I(t-R/c)}{\partial t} + \frac{z^2}{cR^3}\frac{\partial I(t-R/c)}{\partial t} + \frac{z^2}{c^2 R^2}\frac{\partial^2 I(t-R/c)}{\partial t^2}$$

$$\frac{\partial^2 I(t-R/c)}{\partial r\partial z} = \frac{\partial}{\partial r}\left[-\frac{z}{cR}\frac{\partial I(t-R/c)}{\partial t}\right]$$

$$= -\frac{z}{c}\left[\left(-\frac{r}{R^3}\right)\frac{\partial I(t-R/c)}{\partial t} + \frac{1}{R}\frac{\partial}{\partial t}\left(-\frac{r}{cR}\frac{\partial I(t-R/c)}{\partial t}\right)\right]$$

$$= \frac{zr}{cR^3}\frac{\partial I(t-R/c)}{\partial t} + \frac{zr}{c^2 R^2}\frac{\partial^2 I(t-R/c)}{\partial t^2}$$

Assemble the results:

$$\overline{a}_r \frac{\partial}{\partial r}(\nabla \cdot \overline{A}) = \frac{\mu_0 dz'}{4\pi}\left\{\frac{1}{R}\left[\frac{zr}{cR^3}\frac{\partial I(t-R/c)}{\partial t} + \frac{zr}{c^2 R^2}\frac{\partial^2 I(t-R/c)}{\partial t^2}\right]\right.$$

$$+\left(-\frac{r}{R^3}\right)\left[-\frac{z}{cR}\frac{\partial I(t-R/c)}{\partial t}\right] + \left(-\frac{z}{R^3}\right)\left[-\frac{r}{cR}\frac{\partial I(t-R/c)}{\partial t}\right]$$

$$\left. + \frac{3zr}{R^5}I(t-R/c)\right\}\overline{a}_r$$

$$= \frac{\mu_0 dz'}{4\pi}\left[\frac{3rz}{R^5}I(t-R/c) + \frac{3rz}{cR^4}\frac{\partial I(t-R/c)}{\partial t} + \frac{rz}{c^2 R^3}\frac{\partial^2 I(t-R/c)}{\partial t^2}\right]\overline{a}_r$$

$$\overline{a}_z \frac{\partial}{\partial z}(\nabla \cdot \overline{A}) = \frac{\mu_0 dz'}{4\pi}\left\{\frac{1}{R}\left[\left(-\frac{1}{cR}+\frac{z^2}{cR^3}\right)\frac{\partial I(t-R/c)}{\partial t} + \frac{z^2}{c^2 R^2}\frac{\partial^2 I(t-R/c)}{\partial t^2}\right]\right.$$

$$-\frac{z}{R^3}\left[-\frac{z}{cR}\frac{\partial I(t-R/c)}{\partial t}\right] - \frac{1}{R^3}I(t-R/c) + \frac{3z^2}{R^5}I(t-R/c)$$

$$\left. -\frac{z}{R^3}\left[-\frac{z}{cR}\frac{\partial I(t-R/c)}{\partial t}\right]\right\}\overline{a}_z$$

$$= \frac{\mu_0 dz'}{4\pi}\left[\left(\frac{3z^2}{R^5}-\frac{1}{R^3}\right)I(t-R/c) + \left(\frac{3z^2}{cR^4}-\frac{1}{cR^2}\right)\frac{\partial I(t-R/c)}{\partial t}\right.$$

$$\left. + \frac{z^2}{c^2 R^3}\frac{\partial^2 I(t-R/c)}{\partial t^2}\right]\overline{a}_z$$

Thus, the r- and z-components of $\nabla\phi$ become:

r-component

$$\nabla\phi_r = -c^2\int_{-\infty}^{t}\overline{a}_r\frac{\partial}{\partial r}(\nabla\cdot\overline{A})d\tau = -\frac{dz'}{4\pi\varepsilon_0}\left[\frac{3rz}{R^5}\int_{0}^{t}I(\tau-R/c)d\tau\right.$$

$$\left. + \frac{3rz}{cR^4}I(t-R/c) + \frac{rz}{c^2 R^3}\frac{\partial I(t-R/c)}{\partial t}\right]\overline{a}_r$$

$$\nabla\phi_z = -c^2\int_{-\infty}^{t}\overline{a}_z\frac{\partial}{\partial z}(\nabla\cdot\overline{A})d\tau = -\frac{dz'}{4\pi\varepsilon_0}\left[\left(\frac{3z^2}{R^5}-\frac{1}{R^3}\right)\int_{0}^{t}I(\tau-R/c)d\tau\right.$$

z-component

$$\left. + \left(\frac{3z^2}{cR^4}-\frac{1}{cR^2}\right)I(t-R/c) + \frac{z^2}{c^2 R^3}\frac{\partial I(t-R/c)}{\partial t}\right]\overline{a}_z$$

where the lower integration limit is changed from $-\infty$ to zero and c^2 is replaced with $(\mu_0\varepsilon_0)^{-1}$.

Finally, the time derivative of $\overline{A}$ is

$$\frac{\partial \overline{A}}{\partial t} = \frac{\partial}{\partial t}\left[\frac{\mu_0 dz'}{4\pi}\frac{I(t - R/c)}{R}\overline{a}_z\right] = \frac{dz'}{4\pi\varepsilon_0}\left(\frac{1}{c^2 R}\frac{\partial I(t - R/c)}{\partial t}\right)\overline{a}_z \qquad \underline{\text{z-component}}$$

where μ_0 is replaced with $(\varepsilon_0 c^2)^{-1}$.

The electric field intensity expression has three terms and is given by

$$d\overline{E} = -\nabla\phi_r - \nabla\phi_z - \partial\overline{A}/\partial t \qquad (A3.10)$$

Recall that $\nabla\phi_r$, $\nabla\phi_z$, and $\overline{A}$ are for the differential current element at $z' = 0$. One can see from Eq. A3.10 that $d\overline{E}$ has the radial (r) and vertical (z) components and that $\nabla\phi$ contributes to both the r- and z-components of $d\overline{E}$, while $\partial\overline{A}/\partial t$ only to the z-component. The two components of $d\overline{E}$ can be expressed as follows:

$$dE_r = \frac{dz'}{4\pi\varepsilon_0}\left[\frac{3rz}{R^5}\int_0^t I(\tau - R/c)d\tau + \frac{3rz}{cR^4}I(t - R/c) + \frac{rz}{c^2 R^3}\frac{\partial I(t - R/c)}{\partial t}\right] \qquad (A3.11)$$

$$dE_z = \frac{dz'}{4\pi\varepsilon_0}\left[\frac{2z^2 - r^2}{R^5}\int_0^t I(\tau - R/c)d\tau + \frac{2z^2 - r^2}{cR^4}I(t - R/c) - \frac{r^2}{c^2 R^3}\frac{\partial I(t - R/c)}{\partial t}\right]$$

$$(A3.12)$$

In the dE_z equation, the geometrical factors in the three terms were obtained as follows:

$$\frac{3z^2}{R^5} - \frac{1}{R^3} = \frac{3z^2 - R^2}{R^5} = \frac{3z^2 - z^2 - r^2}{R^5} = \frac{2z^2 - r^2}{R^5} \qquad \text{1st term (electrostatic)}$$

$$\frac{3z^2}{cR^4} - \frac{1}{cR^2} = \frac{3z^2 - R^2}{cR^4} = \frac{3z^2 - z^2 - r^2}{cR^4} = \frac{2z^2 - r^2}{cR^4} \qquad \text{2nd term (induction)}$$

$$\frac{z^2}{c^2 R^3} - \frac{1}{c^2 R} = \frac{z^2 - R^2}{c^2 R^3} = \frac{z^2 - z^2 - r^2}{c^2 R^3} = -\frac{r^2}{c^2 R^3} \qquad \text{3rd term (radiation)}$$

A3.2 Elevated differential current element ($z' > 0$) and its image

Now we assume that the lower half-space ($z < 0$) is perfectly conducting ground and that the differential current element is elevated to height $z' > 0$. The field point remains at the same position as in Section A3.1. The presence of perfectly conducting ground can be accounted for by using the image theory; that is, by placing an image current element, having the same magnitude and the same direction as the real one, at distance z' below the ground surface. The vectorial sum of field contributions from the real and image current elements at field point P at $z > 0$ in this configuration (see Fig. A3.1b) is identical to the field at point P in the original configuration (a single current element above the ground plane).

In order to make the equations for $d\overline{B}$ and $d\overline{E}$ derived in Section A3.1 applicable to the real and image current elements located at z' and $-z'$, respectively, the following changes are to be made.

For the real current element we simply replace z with $(z - z')$ in the electric field equations:

$$dE_r = \frac{dz'}{4\pi\varepsilon_0}\left[\frac{3r(z-z')}{R^5}\int_0^t I(z', \tau - R/c)d\tau + \frac{3r(z-z')}{cR^4}I(z', t-R/c)\right.$$
$$\left. + \frac{r(z-z')}{c^2R^3}\frac{\partial I(z', t-R/c)}{\partial t}\right] \qquad (A3.13)$$

$$dE_z = \frac{dz'}{4\pi\varepsilon_0}\left[\frac{2(z-z')^2 - r^2}{R^5}\int_0^t I(z', \tau - R/c)d\tau + \frac{2(z-z')^2 - r^2}{cR^4}I(z', t-R/c)\right.$$
$$\left. - \frac{r^2}{c^2R^3}\frac{\partial I(z', t-R/c)}{\partial t}\right] \qquad (A3.14)$$

where $R = \sqrt{r^2 + (z - z')^2}$. The equation for the ϕ-component of magnetic field (B_ϕ) is similar to Eq. A3.4, the only differences being the change in the argument of current function from $(t-R/c)$, where $z' = 0$ is implied, to $(z', t-R/c)$ and different expression for R (see above):

$$dB_\phi = \frac{\mu_0 dz'}{4\pi}\left[\frac{r}{R^3}I(z', t-R/c) + \frac{r}{cR^2}\frac{\partial I(z', t-R/c)}{\partial t}\right] \qquad (A3.15)$$

Note that the two terms in the brackets in Eq. A3.15 are transposed relative to Eq. A3.4 in order to make the magnetic field equation consistent with the traditional formulation in which the $\partial I/\partial t$ term follows the I term.

Equations for the image current element can be obtained from the equations for the real one (Eqs. A3.13–A3.15 above) by replacing $(z - z')$ with $(z + z')$ and R with $R_I = \sqrt{r^2 + (z + z')^2}$.

A3.3 Elevated differential current element above ground and its image; field point on the ground surface

When $z = 0$ (point P on the ground surface; see Fig. A3.1c), $R_I = R$ and, hence, contributions from the real and image sources are equal to each other. Thus, the effect of perfectly conducting ground plane on the magnetic field on that plane is to double the contribution from the real current element.

Similarly, the contributions from the real and image current elements to E_z are equal, causing the field doubling effect, while the contributions to E_r are equal in magnitude and opposite in sign. Hence, the radial component of electric field on perfectly conducting ground is zero, as required by the boundary condition on the tangential component of electric field on the dielectric/conductor interface; only normal component can exist on the surface of a perfect conductor.

The total electric and magnetic fields at the ground surface ($z = 0$) due to the elevated differential current element and its image are given by

$$dE_z(t) = \frac{dz'}{2\pi\varepsilon_0} \left[\frac{2z'^2 - r^2}{R^5} \int_0^t I(z', \tau - R/c)d\tau + \frac{2z'^2 - r^2}{cR^4} I(z', t - R/c) \right.$$
$$\left. - \frac{r^2}{c^2R^3} \frac{\partial I(z', t - R/c)}{\partial t} \right]$$
(A3.16)

$$dB_\phi(t) = \frac{\mu_0 dz'}{2\pi} \left[\frac{r}{R^3} I(z', t - R/c) + \frac{r}{cR^2} \frac{\partial I(z', t - R/c)}{\partial t} \right]$$
(A3.17)

Equations A3.16 and A3.17 are applicable to an electrically short vertical dipole above ground, such as the compact intracloud discharge (CID; see Appendix 4). The three terms in Eq. A3.16 are named electrostatic, induction, and radiation, and in Eq. A3.17 the two terms are magnetostatic and radiation.

A3.4 Vertical lightning channel above ground; field point on the ground surface

This configuration is usually applied to the return-stroke process in which a current wave propagates from the ground level up along the channel.

Integrating Eq. A3.16 over the radiating channel length H(t) we get

$$E_z(t) = \frac{1}{2\pi\varepsilon_0} \left[\int_0^{H(t)} \frac{2z'^2 - r^2}{R^5} \int_0^t I(z', \tau - R/c)d\tau dz' + \int_0^{H(t)} \frac{2z'^2 - r^2}{cR^4} I(z', t - R/c)dz' \right.$$
$$\left. - \int_0^{H(t)} \frac{r^2}{c^2R^3} \frac{\partial I(z', t - R/c)}{\partial t}dz' \right]$$
(A3.18)

Alternatively, we can write

$$E_z(t) = \frac{1}{2\pi\varepsilon_0} \left[\int_0^{H(t)} \frac{(2 - 3\sin^2\theta)}{R^3} \int_0^t I(z', \tau - R/c)d\tau dz' + \int_0^{H(t)} \frac{(2 - 3\sin^2\theta)}{cR^2} I(z', t - R/c)dz' \right.$$
$$\left. - \int_0^{H(t)} \frac{\sin^2\theta}{c^2R} \frac{\partial I(z', t - R/c)}{\partial t}dz' \right]$$
(A3.19)

where the following relations were used:

$$r^2/R^2 = \sin^2(180° - \theta) = \sin^2\theta$$

$$z'^2/R^2 = \cos^2(180° - \theta) = (-\cos\theta)^2 = \cos^2\theta$$

$$\frac{2z'^2 - r^2}{R^2} = 2\cos^2\theta - \sin^2\theta + 2\sin^2\theta - 2\sin^2\theta = 2 - 3\sin^2\theta$$

Note that the lower integration limit in the time integral here is zero, which is different from that $(z'/v_f + R(z')/c)$ in Eq. 5.7. This difference is generally immaterial, because no current in the channel at height z' is "seen" at point P during the time interval from $t = 0$ to $t = z'/v_f + R(z')/c$, where v_f is the propagation speed of return-stroke front.

Integrating Eq. A3.17 over the radiating channel length $H(t)$, we get

$$B_\phi(t) = \frac{\mu_0}{2\pi}\left[\int_0^{H(t)} \frac{r}{R^3}I(z',\ t - R/c)dz' + \int_0^{H(t)} \frac{r}{cR^2}\frac{\partial I(z',\ t - R/c)}{\partial t}dz'\right] \qquad \text{(A3.20)}$$

Alternatively, using $r/R = \sin(180° - \theta) = \sin\theta$, we can write

$$B_\phi(t) = \frac{\mu_0}{2\pi}\left[\int_0^{H(t)} \frac{\sin\theta}{R^2}I(z',\ t - R/c)dz' + \int_0^{H(t)} \frac{\sin\theta}{cR}\frac{\partial I(z',\ t - R/c)}{\partial t}dz'\right] \qquad \text{(A3.21)}$$

The radiating channel length $H(t)$ in Eqs. A3.18–A3.21 for an upward-moving current wave is found from Eq. 5.9.

Eqs. A3.18–A3.21 are exact, provided that the lightning channel is vertical, the ground is perfectly conducting, and the field point is on the ground surface.

Eqs. A3.16 and A3.17 are applicable to the case of electrically-short (Hertzian) vertical dipole above perfectly conducting ground and the field point at ground level. They can be also used (after integration over z') for an elevated vertical channel of arbitrary length.

Eqs. A3.13–A3.15 (and their counterparts for the image current element) are the basis for deriving field equations for the case of a field point located at an arbitrary position in space, still above perfectly conducting ground. Note that at an elevated field point the electric field will have both z- and r-components, because the inclined distances from the real and image current elements to the field point are different.

Additional reading

Thottappillil, R. (2014). Computation of electromagnetic fields from lightning discharge, in *The Lightning Flash*, 2nd Edition, ed. V. Cooray, London: IEE, 351–403.

Appendix 4 Compact intracloud discharges (CIDs)

As noted in Chapter 1, there is a special type of lightning that is thought to be the most intense natural producer of HF–VHF (3–300 MHz) radiation on Earth. It is referred to as Compact Intracloud Discharge (CID). CIDs received their name due to their relatively small (hundreds of meters) spatial extent (Smith et al., 1999). They tend to occur at high altitudes (mostly above 10 km), appear to be associated with strong convection (however, even the strongest convection does not always produce CIDs), tend to produce less light than other types of lightning discharges, and produce single VLF–LF electric field pulses (Narrow Bipolar Pulses or NBPs) having typical full widths of 10 to 30 μs and amplitudes of the order of 10 V/m at 100 km, which is higher than for return strokes in cloud-to-ground flashes. As an illustration of their VLF–LF intensity, 48 CIDs examined in detail by Nag et al. (2010) were recorded by four to 22 (11 on average) stations of the US National Lightning Detection Network (NLDN), whose average baseline is 300–350 km. According to Nag et al. (2010), the majority (more than 70 percent) of CIDs appeared to occur in isolation from any other lightning processes. Various electromagnetic signatures of a CID are shown in Fig. A4.1. Perhaps the most puzzling feature of these mysterious lightning events is the fact that their VHF radiation (which is indicative of preliminary breakdown and leader processes creating a new lightning channel), and their wideband signature (which is indicative of a current wave propagating along an already existing channel) appear to be generated at the same time, as seen in Fig. A4.1d where those signatures are superimposed.

As seen in Fig. A4.1, there is evidence of reflections (five secondary peaks labeled S1 to S5) in the wideband electric field and particularly dE/dt signatures, although generally they may be undetectable in measured field waveforms. In Florida, 15 percent of CIDs exhibited oscillations that are indicative of reflections. These reflections, most likely occurring at CID channel ends, influence the magnitude of the overall CID electric field waveform and are responsible for its fine structure, as well as, by inference, for "noisiness" of dE/dt waveforms and for accompanying HF–VHF bursts. On the basis of the experimental evidence of multiple reflections and modeling, Nag and Rakov (2010) inferred that, from the electromagnetic point of view, the CID is essentially a bouncing-wave phenomenon. Some tens of reflections may occur at both radiating-channel ends, possibly serving to maintain channel conductivity. The process can be viewed as a long wave repeatedly folding on itself. The bouncing-wave CID model developed by Nag and Rakov (2010) allows reproduction of both the two-station measurements of CID electric

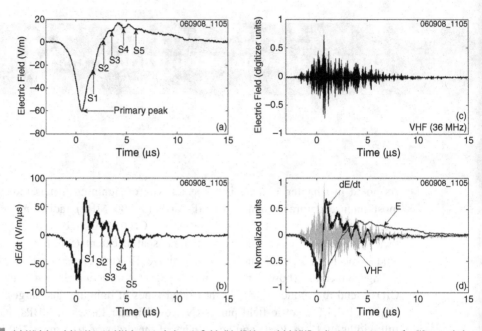

Fig. A4.1 (a) Wideband (16 Hz–10 MHz) vertical electric field, (b) dE/dt, and (c) VHF radiation signatures of a CID recorded at the Lightning Observatory in Gainesville, Florida. It occurred at an unknown distance and transferred negative charge downward. The three signatures are overlaid in (d) for direct comparison. S1–S5 are five secondary peaks appearing as pronounced oscillations in (b) and mostly as shoulders in (a). These secondary peaks are indicative of multiple reflections. Adapted from Nag and Rakov (2010).

fields by Eack (2004) and the ringing CID signatures described above. The question of the formation of CID channel is still open.

CIDs appear to be capable of altering the course of lightning discharge activity on an individual flash scale (Nag et al., 2010) and possibly contributing to production of gigantic jets. Krehbiel et al. (2008) reported a CID that occurred 800 ms prior to a gigantic jet. Also, Lu et al. (2011) reported on two gigantic jets, one in Florida and one in Oklahoma, that were each preceded by an ordinary cloud flash apparently initiated by a CID, the latter occurring about 300 and 500 ms prior to the initiation of gigantic jet, respectively. The specific context in which CIDs occur remains unclear. An example of CID that occurred prior to the cloud-to-ground flash is shown in Fig. A4.2. The top and bottom panels show correlated wideband electric field and VHF radiation records, respectively, both displayed on a 170 ms timescale. In the top panel, PB stands for the preliminary breakdown, and two return-stroke signatures are labeled RS1 and RS2. Note that, while the CID and return-stroke signatures are comparable in magnitude in the top panel, the amplitude of the VHF waveform of CID far exceeds (dwarfs) those of other lightning processes (PB, RS1, and RS2) seen in the bottom panel. It is possible that the CID played a role in facilitating the PB that resulted in the cloud-to-ground discharge.

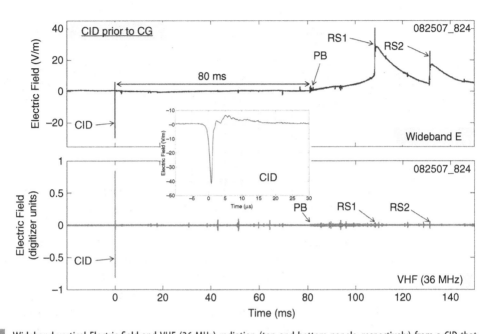

Fig. A4.2   Wideband vertical Electric field and VHF (36 MHz) radiation (top and bottom panels, respectively) from a CID that was followed (80 ms later) by the preliminary breakdown of a multiple-stroke cloud-to-ground discharge. Inset shows the CID signature on an expanded (5 μs per division) timescale. The signals were recorded at the Lightning Observatory in Gainesville, Florida.

Recent efforts in studying CIDs are described in works of Liu et al. (2012), Wang et al. (2012), Jacobson et al. (2012, 2013), Wu et al. (2012, 2013, 2014), Lü et al. (2013), Arabshahi et al. (2014), Cooray et al. (2014), da Silva and Pasko (2015), Karunarathne et al. (2015), and Marshall et al. (2015). Clearly, these mysterious lightning events continue attracting the attention of many research groups.

Appendix 5   Is it true that lightning never strikes the same place twice?

The most common type of lightning discharge between a thundercloud and Earth's surface, which accounts for 90 percent of all lightning discharges involving ground, begins in the cloud. Before the brilliant lightning channel bridges the gap, a lightning process called a "leader" takes place. This process creates a downward branched, low-luminosity channel, not observable with a naked eye or with an ordinary photographic camera. The leader channel extends from the cloud in search for a termination point on the ground. If we consider a terrain that is essentially flat and geologically uniform, the lightning termination point on the ground can be viewed as random. In this case, the saying "lightning never strikes the same place twice" would be essentially true, as illustrated next. A small area of one square meter in an open field in Florida is struck by lightning on average once every 100 millennia. Thus, it is highly unlikely that one can see two lightning strikes to that 1 m^2 area; for all practical purposes this is equivalent to never. It is worth noting that each lightning flash is typically composed of three to five component strokes that usually retrace the same channel to ground. However, these component strokes occur within a second or less and can be detected by a naked eye only as the flickering in luminosity of the lightning channel.

In reality, ground is not homogeneous and, as a result, the descending lightning leader will be attracted to some terrain features more than to others. Grounded metallic objects dominating the area are more likely to be struck by lightning than the surrounding ground or lower structures nearby (this is how lightning rods work). In general, the taller the object, the more often it is struck. For example, a 60 m tower located in Florida is expected to be struck by lightning roughly between once every other year and once every year. As the height of an object increases beyond 100 m or so, a different type of lightning discharge to this object can occur in addition to the type described above. This additional type of lightning also involves a leader process, but the leader channel originates on the object and extends toward the cloud. Clearly, in this case the strike point is predetermined, in contrast with the case of lightning initiated by a descending leader. If the object is very tall, such object-initiated lightning may occur literally dozens of times every thunderstorm season. For example, the Ostankino TV tower in Moscow, Russia, which is 540 m high, experiences about 30 lightning strikes per year. If this tower were relocated to the Tampa Bay area, Florida, where lightning activity is three to four times more intense, it would experience 90 to 120 strikes per year.

In summary, from the lightning standpoint, the saying "lightning never strikes the same place twice" is certainly not true. It does strike its preferential targets such as tall towers over and over again. However, from the standpoint of a small area in a large open field, the saying is essentially true.

Appendix 6 Is it possible to use lightning as an energy source?

It appears to be impractical to utilize lightning energy. Each cloud-to-ground lightning flash involves an energy of roughly 10^9 to 10^{10} J (joules). For comparison, the energy required to operate five 100 W light bulbs continuously for one month is

$$5 \times 100 \text{ W} \times 3600 \text{ s} \times 24 \text{ hr} \times 30 \text{ days} = 1.3 \times 10^9 \text{ J}$$

or about 360 kilowatt hours (1 kW hr = 3.6×10^6 J), which is comparable to the total energy of one lightning flash. Even if it were possible to capture all of a flash's energy (this is not possible since the bulk of this energy is not delivered to the strike point but rather is lost to heating the air and producing thunder, light, and radio waves), one would need to attract 12 flashes to the energy storage facility in order to operate these five light bulbs for one year. The probability of lightning strike to a given point on the ground is very low. For example, a 1 m^2 area in Florida is struck by lightning, on average, once in 10^5 years. A grounded structure protruding above the earth's surface is more likely to be struck by lightning. A 60 m tower located in Florida is expected to be struck by lightning roughly between once every other year and once every year. Thus, one needs 12 to 24 such towers, covering a large area of 1 km^2 or so, to operate five 100 W light bulbs, which is obviously impractical. Most of the US experiences a factor of two to three lower lightning activity than in Florida. As a result, the number of lightning-capturing towers needed to operate only five 100 W bulbs in areas of moderate lightning activity would be 24 to 72. Thus the three main problems with the utilization of lightning energy (leaving aside the issue of energy storage devices) can be formulated as follows:

(1) The power associated with a lightning flash is very high (of the order of 10^{12} W, which is more than ten times greater than the generating capacity of the Three Gorges Dam hydroelectric power station in China), but it is released in pulses of short duration (of the order of 10^{-4} s). As a result, the lightning energy, the integral of the power over the short period of time, is moderate: comparable to the monthly energy consumption of 360 kilowatt hours by five 100 W light bulbs.

(2) Not all the lightning energy is delivered to the strike point. Using a typical value of energy per unit resistance (action integral) of 10^5 A s^2 determined from measurements of the current at the negative lightning channel base and an assumed range of

resistances at the strike point of 10 to 100 Ω, we estimate a range of the lightning energy delivered to the strike point to be from 10^6 to 10^7 J, which is only 10^{-2} to 10^{-4} of the total energy.

(3) The capturing of a sufficiently large number of lightning strikes would require the use of a large number of tall towers, which is impractical.

Appendix 7 Lightning safety

Lightning safety experts (Holle et al., 1999; Walsh et al., 2013) have presented their consensus views of personal safety from lightning that can be summarized as follows:

- No place is absolutely safe from the lightning threat; however, some places are much safer than others.
- Large enclosed structures (substantially constructed buildings) are much safer than smaller or open structures. The risk of lightning injury depends on whether the structure incorporates lightning protection, the construction materials used, and the size of the structure.
- In general, fully enclosed metal vehicles such as cars, trucks, buses, vans, fully enclosed farm vehicles, etc. provide good shelter from lightning.
- One should avoid being outside in the presence of lightning. When the lightning threat develops, go quickly to a substantial building or fully enclosed vehicle.
- When inside a building one should avoid the use of a corded telephone or any contact with conductive paths with exposure to the outside, such as wiring or plumbing.
- Where groups of people are involved, an action plan for getting to a lightning-safe place must be made in advance by the responsible individuals.

Following these relatively simple personal safety rules can considerably reduce the chances of being killed or injured by lightning.

Figure A7.1 shows a triggered-lightning discharge (via a 7 m air gap) to a fully enclosed car with a live rabbit inside. The experiment was carried out on December 20, 1980 at the Kahokugata lightning triggering facility in Japan. The strike terminated on the radio antenna, the top 30 cm of which was melted and vaporized. Also, the strike melted a hole (1 cm in diameter) in the car body near the base of the antenna. However, the rabbit inside the car was not harmed, which confirms that fully enclosed vehicles generally provide good shelter from lightning.

Many lightning strike victims can survive their encounter with lightning, especially with timely medical treatment, including cardiopulmonary resuscitation (CPR). Individuals struck by lightning do not carry an electric charge, so it is safe to touch them, and indeed it is imperative to do so in order to administer CPR.

Further information on locations that are unsafe during thunderstorms is given by Walsh et al. (2013). Their statements and recommendations are quoted below:

"Even though large, substantial buildings containing electrical wiring and plumbing are generally classified as safe, there may still be a potential risk of lightning injury in certain

Fig. A7.1   Triggered-lightning strike to a fully enclosed car with a live rabbit inside. The strike terminated on the radio antenna. The rabbit inside the car was not harmed. Note flashovers across the tires (which were not damaged) and ground surface arcs. Courtesy of S. Sumi.

situations indoors. Lightning can enter a building through electrical or telephone wiring and plumbing, which makes locker-room shower areas, swimming pools (indoor and outdoor), landline telephones, and electrical appliances unsafe during thunderstorms because of the potential contact injury. Even if the building is customarily grounded, lightning is often fast enough and powerful enough to spread and injure someone before the ground faults or other systems are triggered to protect the person touching any of these systems. Indoor swimming pools are just as dangerous as outdoor pools because lighting, heating, plumbing, and drains used in indoor pools ultimately connect to materials outside the building that can be used to transmit the lightning energy into the building or pool. Other areas in substantial buildings, such as a garage with an open door, near open windows, and press boxes with open windows are not safe because the person is not completely within an enclosed area of the building. Rare reports describe people killed or injured by lightning in their homes while talking on hardwired telephones, taking a shower, or standing near household appliances such as dishwashers, stoves, or refrigerators. In the absence of working cell phones, hardwired telephones can still be used cautiously to contact emergency services if needed. One must balance the risk of using the phone to activate EMS for an emergency against the lesser risk of a lightning injury while on the phone for a short time; time spent on a corded telephone should be minimized while lightning is in the area. Further, injury from acoustic damage can occur via loud static from a hardwired telephone earpiece caused by a nearby lightning strike. Many 911 operators, dispatchers, and others using telephones or some car radio communication devices have been injured by lightning, although the injury is usually due to contact electrical injury from lightning energy transmitted from towers through operators' headphones or other hardwired devices and not from acoustic trauma. These injuries have markedly decreased with the advent of wireless systems.

"Because they are not hardwired, cellular and cordless telephones away from their base are completely safe for communication or summoning help during a thunderstorm. It should be also emphasized that metal does not attract lightning, including watches, jewelry, cell phones, and MP3 players. The primary danger of wireless devices is not from attracting lightning but that they may distract individuals from noticing many dangers in their surroundings, including thunder, which is the primary warning that lightning is close enough to be dangerous. Reports of burns caused by the heating of the metal or minor explosion of the batteries and electronics have misled some people to think that lightning struck the device and, thus, that lightning was attracted to the object, but this is an effect and not a cause or an instigator of lightning.

"If people cannot reach a safer location when thunderstorms are in their area, they should at least avoid the riskiest locations and activities, including elevated places, open areas, tall isolated objects, and being in, on, or at the edge of large bodies of water, including swimming pools. One should never seek shelter near or under trees to keep dry during a thunderstorm. It is always much better to go to a safe location."

Timely warning of a lightning threat is crucial for personal safety. If thunderstorms are predicted in the area, it might be better not to plan – or to postpone – outdoor activities such as golfing, boating, and hiking. Dark clouds should generally be viewed as potential lightning producers and as a signal to consider termination of outdoor activities. Any time that visual (flashes) or audible (thunder) indications of a thunderstorm exist, moving in or remaining in reasonably safe shelters, such as large buildings and all-metal-body cars, should be considered. The distance to the lightning can be estimated by counting the number of seconds between the flash and the thunder and dividing the result by three (to obtain kilometers) or by five (to obtain miles). For example, if the time interval between the flash and the thunder is 30 s, the lightning is approximately at a distance of 10 km or six miles. A thunderstorm can cover this distance in 15 minutes or so. However, many lightning discharges strike ground at more than one point, with separation between the strike points (produced within less than a second) being up to several kilometers. An AM radio, tuned to the low end of the broadcast band (550 kHz), acts as a primitive lightning detector by responding with periodic static "crashes" to electromagnetic radiation pulses produced by lightning discharges. The US National Lightning Detection Network (NLDN) continuously provides nationwide information on lightning activity in nearly real time.

Appendix 8   Lightning makes glass

This appendix is based on the Domenic Labino Lecture given by the author of this book at the 1999 Glass Art Society meeting in St. Petersburg, Florida.

A8.1 Introduction

Mother Nature makes glass each time a large amount of energy is released during a sufficient period of time at the Earth's surface, provided that the soil composition is suitable for making glass. The latter condition is satisfied, for example, by sandy soil, with the resultant natural glass being silica glass named "lechatelierite" after the French chemist Henry Le Châtelier (1850–1936). There are two phenomena that are responsible for making natural glass on Earth: meteorites and lightning. Glass that is made as a result of the collision of a meteorite with the Earth's surface is called meteoritic glass or tektite. Glass (a glassy object, to be exact) that is made as a result of a cloud-to-ground lightning discharge is called a fulgurite (from the Latin "fulgur," which means "lightning"). Fulgurites come in a great variety of forms and can be viewed as nature's own works of art. It is worth noting that lechatelierite (natural silica glass) is not present in obsidian, a glass-like material associated with volcanic activity. On the other hand, volcanic activity is known to generate lightning which, if it strikes sandy soil, may produce a fulgurite. Silica glass has been also made as a result of nuclear explosions. In 1945, the first nuclear bomb (equivalent to 18,000 tons of TNT) was detonated in the New Mexico desert. The explosion formed a crater about 700 m in diameter, glazed with a dull gray-green silica glass. This glass was named "trinitite" after Trinity Site where the first nuclear bomb test was conducted.

A8.2 Characterization of lightning

On average, some tens to 100 lightning discharges occur every second on the Earth. Only about one quarter of them involve the ground (others occur in the cloud, between clouds, or between cloud and clear air) and can potentially make fulgurites. Some parts of Florida receive more than 12 lightning strikes per square kilometer per year. This is the highest level of lightning activity in the United States.

Each cloud-to-ground lightning involves energy of roughly 10^9–10^{10} joules. Most of the lightning energy is spent to produce thunder, hot air, light, and radio waves, so that only a small fraction of the total energy is available at the strike point. However, it is well known that this small fraction of the total lightning energy is sufficient to kill people and animals, start fires, and cause considerable mechanical damage to various structures. Lightning is also a major source of electrical disturbances.

The peak temperature of lightning channel is of the order of 30,000 K, which is five times higher than the surface temperature of the Sun (the temperature of the solar interior is 10^7 K). The lightning peak temperature is considerably higher than silica's melting point, which is somewhere between 1600 and 2000°C, depending on moisture content, but whether or not silica sand melts and glass is produced depends, besides other, not-well-understood factors, on lightning duration. Some lightning strokes last (since a contact with ground is made) for less than a millisecond; others linger for a significant fraction of a second. Lightning current peaks are usually of the order of tens of kilo-amperes, but occasionally may exceed 100 kA. The long-lasting current components are typically in the range of tens to hundreds of amperes. The latter are thought to be responsible for making fulgurites.

In the case of natural lightning, it is usually unknown when and where the discharge is going to occur. These uncertainties are largely removed when lightning is artificially initiated (triggered) from an overhead natural thundercloud with the so-called rocket-and-wire technique. Some of the most interesting fulgurites have been created in triggered-lightning experiments.

Some tens of lightning discharges are triggered every year (over 450 to date) at the International Center for Lightning Research and Testing (ICLRT) at Camp Blanding, Florida. The facility is located approximately midway between Jacksonville and Gainesville, Florida, and is used for studying various aspects of atmospheric electricity, lightning, and lightning protection. Many triggered-lightning discharges at Camp Blanding that terminated on ground (as opposed to termination on well-grounded objects or systems) created fulgurites.

A8.3 General information on fulgurites

The earliest discovery of a fulgurite was reportedly made in 1706 by Pastor David Hermann in Germany. Most people have never seen a fulgurite, and if they have they might not have recognized it for what it was. All fulgurites can be divided in two classes: sand fulgurites and rock fulgurites. Sand fulgurites are usually hollow, glass-lined tubes with sand adhering to the outside. Rock fulgurites are formed when lightning strikes the bare surface of rocks. This type of fulgurite appears as thin glassy crust with which may be associated short tubes or perforations lined with glass in the rock. Glass of this type may be relatively low in silica

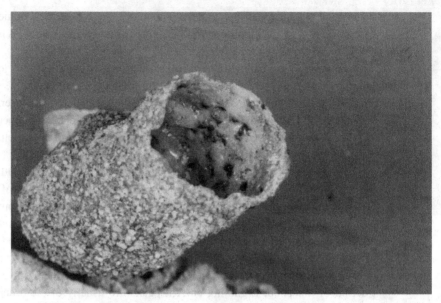

Fig. A8.1   A sand fulgurite made by triggered lightning in 1993 at Camp Blanding, Florida.

and exhibit a wide variety of colors, depending on the composition of the host rock. Rock fulgurites are found on the peaks of mountains.

When lightning strikes sandy soil, the air and moisture present in soil are rapidly heated, and the resultant explosion-like expansion forms the central tubular void. As stated before, quartz sand melts at a temperature of about 1600–2000°C depending on moisture content, and molten glass is pushed to the periphery of the void. Subsequent relatively rapid cooling causes the glass to solidify. A general condition for sand fulgurite formation appears to be the presence of a relatively dry dielectric such as quartz sand overlying a more conducting soil layer or the ground water table, with the depth of the latter probably determining the limit for vertical extent of the fulgurite formation. The diameter of fulgurites ranges from a quarter of an inch to three inches (0.64 to 7.6 cm), and the color varies, depending upon the type of sand from which they were formed. Sand fulgurites are usually tan, grayish, or black; but almost translucent white fulgurites have been found in Florida panhandle beaches. The inner surface is glassy and exhibits numerous bubbles. The walls are usually about 0.5–2 mm thick, but may be paper thin. There appears to be no relation between tube diameter and wall thickness. Sand fulgurites are quite fragile and very difficult to excavate in one piece. An example of sand fulgurite is shown in Fig. A8.1.

Since fulgurites are real glasses, they are very resistant to weathering and are usually well preserved for a long period of time. For this reason they are used as paleoenvironmental indicators. For example, many fulgurites are found in the Sahara Desert, where presently there is little lightning activity, confirming that very different conditions existed in this region in prehistoric times. A fossil fulgurite thought to be 250 million years old has been reported.

Fulgurites have also been produced artificially by passing laboratory arc current through sand. It has been found by researchers at the Technical University of Ilmenau, Germany, that currents higher than 50 kA lasting for some hundreds of microseconds, typical of impulsive components of the lightning current are incapable of making a fulgurite (only some very thin fragments). On the other hand, relatively low magnitude currents of some hundreds of amperes lasting for some hundreds of milliseconds yielded well-formed fulgurites with diameters of 7 to 15 mm. It has been also observed that the higher the current, the larger the cross-sectional dimensions of the fulgurite. Different forms of fulgurites were obtained in dry and wet sand. Fulgurites in wet sand were more curved and had more irregular outer surfaces. The latter feature was attributed to the pressure of vaporized moisture that squashed the fulgurite when the arc pressure in the central tubular void disappeared, while the glass was still plastic.

A8.4 Fulgurites created at Camp Blanding, Florida

A8.4.1 Underground power cable project (1993–4)

In 1993, an experiment, sponsored by Electric Power Research Institute (EPRI), was conducted by Power Technologies, Inc. to study the effects of lightning on underground power cables. In this experiment, three 15 kV coaxial cables with polyethylene insulation between the center conductor and the outer concentric shield (neutral) were buried 5 m apart at a depth of 1 m, and lightning current was injected into the ground at different positions with respect to these cables. One of the cables (Cable A) had an insulting jacket and was placed in PVC conduit, another one (Cable B) had an insulating jacket and was directly buried, and the third one (Cable C) had no jacket and was directly buried. About 20 lightning flashes were triggered directly above the cables which were unenergized. A still photograph of one of those flashes is shown in Fig. A8.2.

The underground power cables were excavated by the University of Florida researchers in 1994. The damage found ranged from minor punctures of the cable jacket to extensive puncturing of the jacket and melting of nearly all the concentric neutral strands near the lightning attachment point. Some damage to the cable insulation was also observed. In the case of the PVC conduit cable installation, the wall of the conduit was melted, distorted and blown open, and the lightning channel had attached to the cable inside and damaged its insulation. Photographs of the damaged parts of the cables are shown in Fig. A8.3.

A total of five fulgurites were found during the excavation of the underground cables. The excavation process was a slow, methodical one and covered an area with dimensions of 4 m × 20 m. Various techniques developed in paleontology were used to remove the fulgurites. The fulgurite excavated over Cable B was nearly vertical

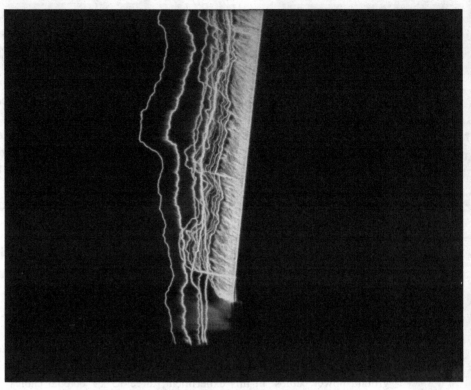

Fig. A8.2 Photograph of a lightning flash triggered in 1993 at Camp Blanding, Florida.

with a length of approximately 1 m and an average diameter of 1.5 cm at the top and about 0.4 cm at the cable. This fulgurite was the most complete fulgurite excavated as part of the underground power cable project. It was unearthed in one piece with very little reconstruction necessary.

A8.4.2 World-record fulgurite (1996)

As noted above, some rocket-triggered discharges strike ground as opposed to terminating on the rocket launcher. Additionally, the Camp Blanding facility, whose area is about 1 square kilometer, receives about five lightning strikes per year that occur naturally, irrespective of lightning-triggering activity. In many cases, the surveillance cameras and observer reports allow researchers to find the strike point on the ground. Such strike points usually appear as holes in the ground with the surrounding grass being killed (as becomes apparent within a few days). When the strike point on the ground is found and flagged, it is impossible to predict if a fulgurite has been created, and, if so, what its shape and dimensions are. One such find in 1996 led to many days of careful digging and resulted in the unearthing of a fulgurite having two mostly vertical branches, one about 16 feet

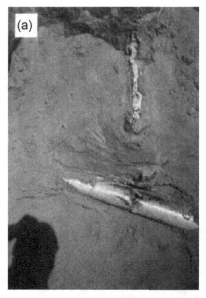

Cable A

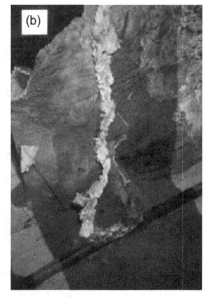

Cable B

Cable C

Fig. A8.3 Lightning damage to underground power cables. (a) coaxial cable in an insulating jacket inside a PVC conduit; note the section of vertical fulgurite in the upper part of the picture (the lower portion of this fulgurite was destroyed during excavation) and the hole melted through the PVC conduit, (b) coaxial cable in an insulating jacket, directly buried; note the fulgurite attached to the cable, (c) coaxial cable whose neutral (shield) was in contact with earth; note that many strands of the neutral are melted through. Photos in (a) and (b) were taken by V. A. Rakov and in (c) by P. P. Barker.

Fig. A8.4 World's longest excavated fulgurite made by triggered lightning in 1996 at Camp Blanding, Florida.

(4.9 m) and the other about 17 feet (5.2 m) long. It was recognized by the Guinness Book of Records as the world's longest excavated fulgurite. The 17 foot branch of the world-record fulgurite is shown in Fig. A8.4.

Appendix 9 Bibliography on triggered lightning experiments and natural lightning observations at Camp Blanding, Florida (1995–2014)

The lightning-triggering facility at Camp Blanding, Florida was established in 1993 by the Electric Power Research Institute (EPRI) and Power Technologies, Inc. (PTI). In September 1994, operation of the facility was transferred to the University of Florida (UF). Over 50 researchers (excluding UF faculty, students, and staff) from 15 countries representing four continents have performed experiments at Camp Blanding concerned with various aspects of atmospheric electricity, lightning, and lightning protection. Since 1995, the Camp Blanding facility has been referred to as the International Center for Lightning Research and Testing (ICLRT) at Camp Blanding, Florida. Presently, the ICLRT is jointly operated by UF and Florida Institute of Technology (FIT) and additionally includes the Lightning Observatory in Gainesville (LOG). Over the 22-year period (1993–2014), the total number of flashes triggered at Camp Blanding was 456; that is, on average about 21 per year, with about 15 (71 percent) of them containing return strokes. Out of the total of 456 flashes, 448 transported negative charge and eight either positive charge or both negative and positive charges to ground. Besides triggered-lightning flashes, discharges occurring naturally over and in the vicinity of the Camp Blanding facility were also studied.

A list of journal papers, in reverse chronological order, presenting or using the results of triggered lightning experiments and natural lightning observations at Camp Blanding, Florida, is given below. The first journal paper was published in 1995, and the total number of papers published in 20 years (1995–2014) is 114. The experiments at Camp Blanding and associated modeling continue, and, hence, the bibliography will be in need of updating soon.

2014

- Simultaneously measured lightning return stroke channel-base current and luminosity, Geophys. Res. Lett., 41, 7799–805, 2014, F. L. Carvalho, D. M. Jordan, M. A. Uman, T. Ngin, W. R. Gamerota, and J. T. Pilkey, doi:10.1002/2014GL062190
- Calculation of current distribution in the lightning protective system of a residential house, IEEE Trans. on Magnetics, Vol. 50, No. 2, 7005404, February 2014, P. Wang, L. Li, and V. A. Rakov, doi: 10.1109/TMAG.2013.2283257
- Evaluation of the GLD360 performance characteristics using rocket-and-wire triggered lightning data, Geophys. Res. Lett., 41, 3636–42, S. Mallick, V. A. Rakov, T. Ngin, W. R. Gamerota, J. T. Pilkey, J. D. Hill, M. A. Uman, D. M. Jordan, A. Nag, and R. K. Said, doi:10.1002/2014GL059920

- Studies of the interaction of rocket-and-wire triggered lightning with various objects and systems (Part 2), Electricity, No. 6, 2014, 4–10, V. A. Rakov
- Studies of the interaction of rocket-and-wire triggered lightning with various objects and systems (Part 1), Electricity, No. 5, 2014, 4–11, V. A. Rakov
- Lightning Observatory in Gainesville (LOG), Florida: A Review of Recent Results, Electric Power Systems Research, 2014, 113, 95–103, V. A. Rakov, S. Mallick, A. Nag, and V. B. Somu, doi: 10.1016/j.epsr.2014.02.037
- Performance characteristics of the NLDN for return strokes and pulses superimposed on steady currents, based on rocket-triggered lightning data acquired in florida in 2004– 2012, J. Geophys. Res.: Atmos., Vol. 119, No. 7, 3825–56, April 16, 2014, S. Mallick, V. A. Rakov, J. D. Hill, T. Ngin, W. R. Gamerota, J. T. Pilkey, C. J. Biagi, D. M. Jordan, M. A. Uman, J. A. Cramer, and A. Nag, doi:10.1002/2013JD021401
- Lightning attachment processes of an "Anomalous" triggered lightning discharge, J. Geophys. Res., 119, 2014, D. Wang, W. R. Gamerota, M. A. Uman, N. Takagi, J. D. Hill, J. Pilkey, T. Ngin, D. M. Jordan, S. Mallick, and V. A. Rakov, doi: 10.1002/ 2013JD020787
- On remote measurements of lightning return stroke peak currents, Atmos. Res., Vols. 135–6, 2014, 306–13, S. Mallick, V. A. Rakov, D. Tsalikis, A. Nag, C. Biagi, D. Hill, D. M. Jordan, M. A. Uman, and J. A. Cramer
- Search for neutrons associated with lightning discharges, the Smithsonian/NASA Astrophysics Data System, published October 7, 2014, J. E. Grove, W. N. Johnson, B. F. Philips, E. A. Wulf, A. L. Hutcheson, L. J. Mitchell, R. S. Woolf, M. M. Schaal, M. A. Uman, D. M. Jordan, and J. R. Dwyer
- Electric field derivative waveforms from dart-stepped-leader steps in triggered lightning, J. Geophys. Res., Atmos., 119, published September 18, 2014, W. R. Gamerota, M. A. Uman, J. D. Hill, T. Ngin, J. Pilkey, and D. M. Jordan, doi:10.1002/2014JD021919
- Does the lightning current go to Zero between ground strokes? Is there a "Current Cutoff"?, Geophysical Research Letters, 41, 3266–73, May 7, 2014, T. Ngin, M. A. Uman, J. D. Hill, R. C. Olsen III, J. T. Pilkey, W. R. Gamerota, and D. M. Jordan, doi:10.1002/2014GL059601
- Negative leader step mechanisms observed in altitude triggered lightning, J. Geophys. Res., Atmospheres, April 10, 2014, C. J. Biagi, M. A. Uman, J. D. Hill, and D. M. Jordan, doi: 10.1002/2013JD020281
- Dart-stepped-leader step formation in triggered lightning, Geophysical Research Letters, 41, 2204–11, published March 24, 2014, W. R. Gamerota, V. P. Idone, M. A. Uman, T. Ngin, J. T. Pilkey, and D. M. Jordan, doi:10.1002/2014GL059627
- The structure of X-ray emissions from triggered lightning leaders measured by a pinhole-type X-ray camera, J. Geophys. Res., published January 30, 2014, M. M. Schaal, J. R. Dwyer, S. Arabshahi, E. S. Cramer, R. J. Lucia, N. Y. Liu, H. K. Rassoul, D. M. Smith, J. W. Matten, A. G. Reid, J. D. Hill, D. M. Jordan, and M. A. Uman, doi:10.1002/ 2013JD020266
- The Physics of Lightning, Physics Reports, 534 (2014), 147–241, Elsevier Press, published January 13, 2014, J. R. Dwyer and M. A. Uman, doi:/10.1016/j.physrep.2013.09.004

2013

- Lightning parameters for engineering applications, Electra, No. 269, August 2013, 58–65, V. A. Rakov, A. Borghetti, C. Bouquegneau, W. A. Chisholm, V. Cooray, K. Cummins, G. Diendorfer, F. Heidler, A. Hussein, M. Ishii, C. A. Nucci, A. Piantini, O. Pinto, Jr., X. Qie, F. Rachidi, M. M. F. Saba, T. Shindo, W. Schulz, R. Thottappillil, S. Visacro, and W. Zischank
- Rocket-and-Wire Triggered Lightning in 2012 Tropical storm debby in the absence of natural lightning, J. Geophys. Res., Atmos, published December 3, 2013, J. T. Pilkey, M. A. Uman, J. D. Hill, T. Ngin, W. R. Gamerota, D. M. Jordan, W. Rison, P. R. Krehbiel, H. E. Edens, M. I. Biggerstaff, and P. Hyland, doi:10.1002/2013JD020501
- Electromagnetic methods of lightning detection, surveys in geophysics (Springer), Vol. 34, No. 6, November 2013, 731–53, V. A. Rakov, doi: 10.1007/s10712–013–9251–1
- The angular distribution of energetic electron and X-ray emissions from triggered lightning leaders," J. Geophys. Res. Atmos., Vol. 118, 11,712–26, October 23, 2013, M. M. Schaal, J. R. Dwyer, H. K. Rassoul, J. D. Hill, D. M. Jordan, and M. A. Uman, doi:10.1002/2013JD019619
- Measurement and analysis of ground-level electric fields and wire-base current During the rocket-and-wire lightning triggering process, J. Geophys. Res. Atmos., Vol. 118, 10,041–55, published September 10, 2013, T. Ngin, M. A. Uman, J. D. Hill, J. Pilkey, W. R. Gamerota, D. M. Jordan, and R. C. Olsen, doi:10.1002/jgrd.50774
- Initiation processes of return strokes in rocket-triggered lightning, J. Geophysical Research: Atmospheres, J. Geophys. Res., Vol. 118, 9880–8, published September 6, 2013, D. Wang, N. Takagi, W. R. Gamerota, M. A. Uman, J. D. Hill, and D. M. Jordan, doi:10.1002/jgrd.50766
- The physics of lightning, surveys in geophysics (Springer), 2013, Vol. 34, No. 6, November 2013, 701–29, V. A. Rakov, doi: 10.1007/s10712–013–9230–6
- Correlated lightning mapping array and radar observations of the initial stage of three sequentially triggered Florida lightning discharges, J. Geophys. Res., Vol. 118, 8460–81, published August 15, 2013, J. D. Hill, J. Pilkey, M. A. Uman, D. M. Jordan, W. Rison, P. R. Krebhiel, M. I. Biggerstaff, P. Hyland, and R. Blakeslee, doi:10.1002/jgrd.50660
- Review of recent developments in lightning channel corona sheath research, Atmos. Res., Vols. 129–30, 2013, 117–22, G. Maslowski and V. A. Rakov, doi:10.1016/j.atmosres.2012.05.028
- An "Anomalous" triggered lightning flash in Florida, J. Geophys. Res., Vol. 118, 3402 14, published April 25, 2013, W. R. Gamerota, M. A. Uman, J. D. Hill, J. Pilkey, T. Ngin, D. M. Jordan, and C. T. Mata, doi: 10.1002/jgrd.50261

2012

- Geometrical and electrical characteristics of the initial stage in Florida triggered lightning, Geophys. Res. Lett., Vol. 39, L09807, 2012, J. D. Hill, J. Pilkey, M. A. Uman, D. M. Jordan, W. Rison, and P. R. Krehbiel, doi:10.1029/2012GL051932

- Observation of a gamma-ray flash at ground level in association with a cloud-to-ground lightning return stroke, J. Geophys. Res., Vol. 117, A10303, 2012, J. R. Dwyer, M. M. Schaal, E. Cramer, S. Arabshahi, N. Liu, H. K. Rassoul, J. D. Hill, D. M. Jordan, and M. A. Uman, doi:10.1029/2012JA017810
- "Chaotic" dart leaders in triggered lightning: Electric fields, X-rays, and source locations, J. Geophys. Res., Vol. 117, D03118, 2012, J. D. Hill, M. A. Uman, D. M. Jordan, J. R. Dwyer, and H. Rassoul, doi:10.1029/2011JD016737
- Spatial and energy distributions of X-ray emissions from Leaders in natural and rocket triggered lightning, J. Geophys. Res., Vol. 117, D15201, 2012, M. M. Schaal, J. R. Dwyer, Z. H. Saleh, H. K. Rassoul, J. D. Hill, D. M. Jordan, and M. A. Uman, doi:10.1029/2012JD017897
- Transient current pulses in rocket-extended wires used to trigger lightning, J. Geophys. Res., Vol. 117, D07205, 2012, C. J. Biagi, M. A. Uman, J. D. Hill, V. A. Rakov, and D. M. Jordan, doi:10.1029/2011JD016161
- Characteristics of the initial rising portion of near and far lightning return stroke electric field waveforms, Atmos. Res., 117, 2012, 71–77, A. Nag, V. A. Rakov, D. Tsalikis, J. Howard, C. J. Biagi, D. Hill, M. A. Uman, and D. M. Jordan
- High frequency earthing impedance measurements at Camp Blanding, Florida, Electric Power Systems Research, 85, 50–8, 2012, A. Rousseau, M. Guthrie, and V. Rakov
- The initial stage processes of rocket-and-wire triggered lightning as observed by VHF interferometry, J. Geophys. Res., Vol. 117, D09119, 2012, S. Yoshida, C. J. Biagi, V. A. Rakov, J. D. Hill, M. V. Stapleton, D. M. Jordan, M. A. Uman, T. Morimoto, T. Ushio, Z. I. Kawasaki, and M. Akita, doi:10.1029/2012JD017657
- Experimental investigation and numerical modeling of surge currents in lightning protection system of a small residential structure, Journal of Lightning Research, 2012, 4 (Suppl 1: M4), 18–26, G. Maslowski, S. Wyderka, V. A. Rakov, B. A. DeCarlo, L. Li, J. Bajorek, and R. Ziemba

2011

- High-speed X-ray images of triggered lightning dart leaders, J. Geophys. Res., Vol. 116, D20208, 2011, J. R. Dwyer, M. Schaal, H. K. Rassoul, M. A. Uman, D. M. Jordan, and D. Hill, doi:10.1029/2011JD015973
- High-speed video observations of a lightning stepped leader, J. Geophys. Res., Vol. 116, D16117, 2011, J. D. Hill, M. A. Uman, and D. M. Jordan, doi:10.1029/2011JD015818
- Observations of the initial, upward-propagating, positive leader steps in a rocket-and-wire triggered lightning discharge, Geophys. Res. Lett., Vol. 38, 2011, C. J. Biagi, M. A. Uman, J. D. Hill and D. M. Jordan, doi:10.1029/2011GL049944
- Some inferences from radial electric fields measured inside the lightning-channel corona sheath, IEEE Trans. on EMC, Vol. 53, No. 2, 390–4, 2011, G. Maslowski, V. A. Rakov, and M. Miki
- Evaluation of NLDN performance characteristics Using rocket-triggered lightning data acquired in 2004–2009, J. Geophys. Res., 116, D02123, 2011, A. Nag, S. Mallick,

V. A. Rakov, J. S. Howard, C. J. Biagi, J. D. Hill, M. A. Uman, D. M. Jordan, K. J. Rambo, J. E. Jerauld, B. A. DeCarlo, K. L. Cummins, and J. A. Cramer, doi:10.1029/2010JD014929
- Measured close lightning leader-step electric field-derivative waveforms, J. Geophys. Res., 116, D08201, 2011, J. Howard, M. A. Uman, C. Biagi, D. Hill, V. A. Rakov, and D. M. Jordan, doi:10.1029/2010JD015249
- Determination of the electric field intensity and space charge density versus height prior to triggered lightning, J. Geophys. Res., 116, D15201, 2011, C. J. Biagi, M. A. Uman, J. Gopalakrishnan, J. D. Hill, V. A. Rakov, T. Ngin, and D. M. Jordan, doi:10.1029/2011JD015710
- Simulation of corona at lightning-triggering wire: Current, charge transfer, and the field-reduction effect, J. Geophys. Res., 116, D21115, 2011, Y. Baba and V. A. Rakov, doi:10.1029/2011JD016341

2010

- Observations of stepping mechanisms in a rocket-and-wire triggered lightning flash, J. Geophys. Res., 115, D23215, 2010, C. J. Biagi, M. A. Uman, J. D. Hill, D. M. Jordan, V. A. Rakov, and J. Dwyer, doi:10.1029/2010JD014616
- Attempts to create ball lightning with triggered lightning, J. Atmospheric and Solar-Terrestrial Physics, ATP3151, 2010, J. D. Hill, M. A. Uman, M. Stapleton, D. M. Jordan, A. Chebaro, and C. J. Biagi, doi:10.1016/j.jastp.2010.04.009
- Parameters of rocket-triggered lightning, International Journal of Plasma Environmental Science and Technology (IJPEST), Vol. 4, No. 1, March 2010, 80–5, V. A. Rakov
- Three-dimensional imaging of upward positive leaders in triggered lightning using VHF broadband digital interferometers, Geophys. Res. Lett., 37, L05805, 2010, S. Yoshida, C. J. Biagi, V. A. Rakov, J. D. Hill, M. V. Stapleton, D. M. Jordan, M. A. Uman, T. Morimoto, T. Ushio, and Z.-I. Kawasaki, doi:10.1029/2009GL042065
- Return stroke peak current vs. charge transfer in rocket-triggered lightning, J. Geophys. Res., 115, D12107, 2010, J. Schoene, M. A. Uman, and V. A. Rakov, doi:10.1029/2009JD013066
- RF and X-ray source locations during the lightning attachment process, J. Geophys. Res., 115, D06204, 2010, J. Howard, M. A. Uman, C. J. Biagi, J. D. Hill, J. Jerauld, V. A. Rakov, J. Dwyer, Z. Saleh, and H. Rassoul, doi:10.1029/2009JD012055

2009

- High-speed video observations of rocket-and-wire initiated lightning, Geophys. Res. Let., Vol. 36, L15801, 2009, C. J. Biagi, D. M. Jordan, M. A. Uman, J. D. Hill, W. H. Beasley, and J. Howard, doi:10.1029/2009GL038525
- Properties of the X-ray emission from rocket-triggered lightning as measured by the Thunderstorm Energetic Radiation Array (TERA), J. Geophys. Res., Vol. 114, D17210,

2009, Z. Saleh, J. Dwyer, J. Howard, M. Uman, M. Bakhtiari, D. Concha, M. Stapleton, D. Hill, C. Biagi, and H. Rassoul, doi:10.1029/2008JDO11618

- Overview of recent progress in lightning research and lightning protection, IEEE Trans. on EMC, Special Issue on Lightning, Vol. 51, No. 3, 428–42, August 2009, V. A. Rakov and F. Rachidi
- Measured electric and magnetic fields from an unusual cloud-to-ground lightning flash containing two positive strokes followed by four negative strokes, J. Geophys. Res., 114, D19115, 2009, J. E. Jerauld, M. A. Uman, V. A. Rakov, K. J. Rambo, D. M. Jordan, and G. H. Schnetzer, doi:10.1029/2008JD011660
- Lightning currents flowing in the soil and entering a test power distribution line via its grounding, IEEE Trans. on Power Delivery, Vol. 24, No. 3, July 2009, 1095–103, J. Schoene, M. A. Uman, V. A. Rakov, J. Jerauld, K. J. Rambo, D. M. Jordan, G. H. Schnetzer, M. Paolone, C. A. Nucci, E. Petrache, and F. Rachidi
- Characterization of return stroke currents in rocket-triggered lightning, J. Geophys. Res., 114, D03106, 2009, J. Schoene, M. A. Uman, V. A. Rakov, K. J. Rambo, J. Jerauld, C. T. Mata, A. G. Mata, D. M. Jordan, and G. H. Schnetzer, doi:10.1029/2008JD009873

2008

- Co-location of lightning leader X-ray and electric field change sources, Geophys. Res. Lett., Vol. 35, L13817, 2008, J. Howard, M. A. Uman, J. R. Dwyer, D. Hill, C. Biagi, Z. Saleh, J. Jerauld, and H. K. Rassoul, doi:10.1029/2008GL034134
- Testing of lightning protective system of a residential structure: Comparison of data obtained in rocket-triggered lightning and current surge generator experiments, High-Voltage Engineering Journal, Vol. 34, No. 12, 2575–82, December 2008, G. Maslowski, V. A. Rakov, S. Wyderka, J. Bajorek, B. A. DeCarlo, J. Jerauld, G. H. Schnetzer, J. Schoene, M. A. Uman, K. J. Rambo, D. M. Jordan, and W. Krata
- Characteristics of the optical pulses associated with a downward branched stepped leader, J. Geophys. Res., 113, D21206, W. Lu, D. Wang, N. Takagi, V. Rakov, M. Uman, and M. Miki, doi:10.1029/2008JD010231
- Electric and magnetic fields and field derivatives from lightning stepped leaders and First return strokes measured at distances from 100 to 1000 m, J. Geophys. Res., 113, D17111, 2008, J. Jerauld, M. A. Uman, V. A. Rakov, K. J. Rambo, D. M. Jordan, and G. H. Schnetzer, doi:10.1029/2008JD010171
- Responses of airport runway lighting system to direct lightning strikes: Comparisons of TLM predictions with experimental data, IEEE Trans. on EMC, Vol. 50, No. 3, August 2008, 660–8, N. Theethayi, V. Rakov, and R. Thottappillil
- Distribution of currents in the lightning protective system of a residential building: I. Triggered-lightning experiments, IEEE Trans. on Power Delivery, Vol. 23, No. 4, October 2008, 2439–46, B. A. DeCarlo, V. A. Rakov, J. Jerauld, G. H. Schnetzer, J. Schoene, M. A. Uman, K. J. Rambo, V. Kodali, D. M. Jordan, G. Maxwell, S. Humeniuk, and M. Morgan

- Distribution of currents in the lightning protective system of a residential building: II. Numerical modeling, IEEE Trans. on Power Delivery, Vol. 23, No. 4, October 2008, 2447–55, L. Li and V. A. Rakov
- Experimental study of lightning-induced currents in a buried loop conductor and a grounded vertical conductor, IEEE Trans. on EMC, Vol. 50, No. 1, February 2008, 110–17, J. Schoene, M. A. Uman, V. A. Rakov, J. Jerauld, B. D. Hanley, K.J. Rambo, J. Howard, and B. DeCarlo

2007

- Insights into the ground attachment process of natural lightning gained from an unusual triggered-lightning stroke, J. Geophys. Res., 112, D13113, 2007, J. Jerauld, M. A. Uman, V. A. Rakov, K. J. Rambo, and G. H. Schnetzer, doi:10.1029/2006JD007682
- Lightning return stroke speed, J. Lightning Research, 2007, Vol. 1, 80–9, V. A. Rakov
- Measurements of NOx produced by rocket-triggered lightning, Geophys. Res. Lett., 34, L03816, 2007, M. Rahman, V. Cooray, V. A. Rakov, M. A. Uman, P. Liyanage, B. A. DeCarlo, J. Jerauld, and R. C. Olsen III, doi:10.1029/2006GL027956
- Direct lightning strikes to test power distribution lines—Part I: Experiment and overall results, IEEE Trans. on Power Delivery, Vol. 22, No. 4, October 2007, 2245–53, J. Schoene, M. A. Uman, V. A. Rakov, A. G. Mata, C. T. Mata, K. J. Rambo, J. Jerauld, D. M. Jordan, and G. H. Schnetzer
- Direct lightning strikes to test power distribution lines—Part II: Measured and modeled current division among multiple arresters and grounds, IEEE Trans. on Power Delivery, Vol. 22, No. 4, October 2007, 2236–44, J. Schoene, M. A. Uman, V. A. Rakov, A. G. Mata, C. T. Mata, K. J. Rambo, J. Jerauld, D. M. Jordan, and G. H. Schnetzer
- The lightning striking distance—revisited, J. Electrostatics, 65, 296–306, 2007, V. Cooray, V. Rakov, and N. Theethayi
- Lightning-induced currents in buried coaxial cables: A frequency-domain approach and its validation using rocket-triggered lightning, J. Electrostatics, 65, 322–8, 2007, E. Petrache, M. Paolone, F. Rachidi, C. A. Nucci, V. Rakov, M. Uman, D. Jordan, K. Rambo, J. Jerauld, M. Nyffeler, and J. Schoene

2006

- Estimation of input energy in rocket-triggered lightning, Geophys. Res. Lett., 33, L05702, 2006, V. Jayakumar, V. A. Rakov, M. Miki, M. A. Uman, G. H. Schnetzer, and K. J. Rambo, doi:10.1029/2005GL025141
- Leader/return-stroke-like processes in the initial stage of rocket-triggered lightning, J. Geophys. Res., 111, D13202, 2006, 11 p., R. C. Olsen, V. A. Rakov, D. M. Jordan, J. Jerauld, M. A. Uman, and K. J. Rambo, doi:10.1029/2005JD006790

2005

- An evaluation of the performance characteristics of the U.S. National Lightning Detection Network in Florida using triggered lightning, J. Geophys. Res., 110, D19106, 2005, J. Jerauld, V. A. Rakov, M. A. Uman, K. J. Rambo, D. M. Jordan, K. L. Cummins, and J. A. Cramer, doi:10.1029/2005JD005924
- X-ray bursts associated with leader steps in cloud-to-ground lightning, Geophys. Res. Lett., 32, L01803, 2005, J. R. Dwyer, H. K. Rassoul, M. Al-Dayeh, L. Caraway, A. Chrest, B. Wright, E. Kozak, J. Jerauld, M. A. Uman, V. A. Rakov, D. M. Jordan, and K. J. Rambo, doi:10.1029/2004GL021782
- Close electric field signatures of dart leader/return stroke sequences in rocket-triggered lightning showing residual fields, J. Geophys. Res., 110, D07205, 2005, V. A. Rakov, V. Kodali, D. E. Crawford, J. Schoene, M. A. Uman, K. J. Rambo, and G. H. Schnetzer, doi:10.1029/2004JD005417
- Oxide reduction during triggered-lightning fulgurite formation, J. Atmos. Solar-Ter. Phys., 67, 423–8, 2005, B. E. Jones, K. S. Jones, K. J. Rambo, V. A. Rakov, J. Jerauld, and M. A. Uman
- Lightning-induced disturbances in buried cables – Part II: Experiment and model validation, IEEE Trans. on EMC, Vol. 47, No. 3, 509–20, August 2005, M. Paolone, E. Petrache, F. Rachidi, C. A. Nucci, V. A. Rakov, M. A. Uman, D. Jordan, K. Rambo, J. Jerauld, M. Nyffeler, and J. Schoene
- Initial stage in lightning initiated from tall objects and in rocket-triggered lightning, J. Geophys. Res., 110, D02109, 2005, M. Miki, V. A. Rakov, T. Shindo, G. Diendorfer, M. Mair, F. Heidler, W. Zischank, M. A. Uman, R. Thottappillil, and D. Wang, doi:10.1029/2003JD004474
- A review of ten years of triggered-lightning experiments at Camp Blanding, Florida, Atmos. Res., Vol. 76, No. 1–4, 504–18, 2005, V. A. Rakov, M. A. Uman, and K. J. Rambo
- Triggered-lightning properties inferred from measured currents and very close electric fields, Atmos. Res., Vol. 76, No. 1–4, 355–76, 2005, V. Kodali, V. A. Rakov, M. A. Uman, K. J. Rambo, G. H. Schnetzer, J. Schoene, and J. Jerauld
- A comparison of channel-base currents and optical signals for rocket-triggered lightning strokes, Atmos. Res., Vol. 76, No. 1–4, 412–22, 2005, D. Wang, N. Takagi, T. Watanabe, V. A. Rakov, M. A. Uman, K. J. Rambo, and M. V. Stapleton

2004

- Observed one-dimensional return stroke propagation speeds in the bottom 170 m of a rocket-triggered lightning channel, Geophys. Res. Lett., 31, L16107, 2004, R. C. Olsen, D. M. Jordan, V. A. Rakov, M. A. Uman, and N. Grimes, doi: 10.1029/2004GL020187
- A triggered lightning flash containing both negative and positive strokes, Geophys. Res. Lett., 31, L08104, 2004, J. Jerauld, M. A. Uman, V. A. Rakov, K. J. Rambo, and D. M. Jordan, doi:10.1029/2004GL019457

- Triggered lightning testing of an airport runway lighting system, IEEE Transactions on Electromagnetic Compatibility, Vol. 46, No. 1, 2004, M. Bejleri, V. A. Rakov, M. A. Uman, K. J. Rambo, C. T. Mata, and M. I. Fernandez
- Measurements of X-ray emission from rocket-triggered lightning, Geophysical Research Letters, Vol. 31, L05118, 2004, J. R. Dwyer, H. K. Rassoul, M. Al-Dayeh, L. Caraway, B. Wright, A. Chrest, M. A. Uman, V. A. Rakov, K. J. Rambo, D. M. Jordan, J. Jerauld, and C. Smyth, doi:10.1029/2003GL018770
- A ground level gamma-ray burst observed in association with rocket-triggered lightning, Geophysical Research Letters, Vol. 31, L05119, 2004, J. R. Dwyer, H. K. Rassoul, M. Al-Dayeh, L. Caraway, B. Wright, A. Chrest, M. A. Uman, V. A. Rakov, K. J. Rambo, D. M. Jordan, J. Jerauld, and C. Smyth, doi:10.1029/2003GL018771

2003

- Test of the transmission line model and the traveling current source model with triggered lightning return strokes at very close range, J. Geophys. Res., Vol. 108, No. D23, 4737, J. Schoene, M. A. Uman, V. A. Rakov, K. J. Rambo, J. Jerauld, and G. H. Schnetzer, doi:10.1029/2003JD003683
- Cutoff and re-establishment of current in rocket-triggered lightning, J. Geophys. Res., Vol. 108, No. D23, 4747, V. A. Rakov, D. E. Crawford, V. Kodali, V. P. Idone, M. A. Uman, G. H. Schnetzer, and K. J. Rambo, doi:10.1029/2003JD003694
- Energetic radiation produced by rocket-triggered lightning, Science, 299, 694–7, 2003, J. R. Dwyer, M. A. Uman, H. K. Rassoul, M. Al-Dayeh, E. L. Caraway, J. Jerauld, V. A. Rakov, D. M. Jordan, K. J. Rambo, V. Corbin, and B. Wright
- Statistical characteristics of the electric and magnetic fields and their time derivatives 15 m and 30 m from triggered lightning, J. Geophys, Res, Vol. 108, No. D6, 4192, 2003, J. Schoene, M. A. Uman, V. A. Rakov, V. Kodali, K. J. Rambo, and G. H. Schnetzer, doi:10.1029/2002JD002698
- Measurement of the division of lightning return stroke current among the multiple arresters and grounds of a power distribution line, IEEE Trans. on Power Delivery, Vol. 18, No. 4, 1203–8, 2003, C. T. Mata, V. A. Rakov, K. J. Rambo, P. Diaz, R. Rey, and M. A. Uman

2002

- Close lightning electromagnetic environment for aircraft testing, SAE 2001 Transactions – Journal of Aerospace, 312–19, M. A. Uman, V. A. Rakov, J. Schoene, K. J. Rambo, J. Jerauld, and G. H. Schnetzer
- Surges superimposed on continuing currents in lightning discharges, SAE 2001 Transactions – Journal of Aerospace, 380–5, V. A. Rakov
- Correlated time derivatives of current, electric field intensity, and magnetic Flux density for triggered lightning at 15 m, J. Geophys. Res., 107(D13), art. no. 4160,

2002, 11 p., M. A. Uman, J. Schoene, V. A. Rakov, K. J. Rambo and G. H. Schnetzer, doi:10.1029/2000JD000249

- Electric fields near triggered lightning channels measured with Pockels sensors, J. Geophys. Res., 107(D16), 2002, 11 p., M. Miki, V. A. Rakov, K. J. Rambo, G. H. Schnetzer and M. A. Uman, doi:10.1029/2001JD001087
- Direct lightning strikes to the lightning protective system of a residential building: Triggered-lightning experiments, IEEE Trans. on Power Delivery, 575–86, 2002, V. A. Rakov, M. A. Uman, M. I. Fernandez, C. T. Mata, K. J. Rambo, M. V. Stapleton, and R. R. Sutil

2001

- M-component mode of charge transfer to ground in lightning discharges, J. Geophys. Res., 106, 22,817–31, 2001, V. A. Rakov, D. E. Crawford, K. J. Rambo, G. H. Schnetzer, M. A. Uman, and R. Thottappillil
- The close lightning electromagnetic environment: Dart-leader electric field change versus distance, J. Geophys. Res., 106, 14,909–17, 2001, D. E. Crawford, V. A. Rakov, M. A. Uman, G. H. Schnetzer, K. J. Rambo, M. V. Stapleton, and R. J. Fisher

2000

- Luminosity waves in branched channels of two negative lightning flashes, Journal of Atmospheric Electricity, 20, 91–7, 2000, D. Wang, N. Takagi, T. Watanabe, V. A. Rakov, and M. A. Uman
- Time derivative of the electric field 10, 14, and 30 m from triggered lightning strokes, J. Geophys. Res., 105, 15,577–95, 2000, M. A. Uman, V. A. Rakov, G. H. Schnetzer, K. J. Rambo, D. E. Crawford and R. J. Fisher
- EMTP modeling of a triggered-lightning strike to the phase conductor of an overhead distribution line, IEEE Trans. on Power Delivery, Vol. 15, No. 4, 1175–81, 2000, C. T. Mata, M. I. Fernandez, V. A. Rakov, and M. A. Uman

1999

- Lightning makes glass, Journal of the Glass Art Society, 45–50, 1999, V. A. Rakov
- Observed leader and return-stroke propagation characteristics in the bottom 400 m of the rocket triggered lightning channel, J. Geophys. Res., 104, 14,369–76, 1999, D. Wang, N. Takagi, T. Watanabe, V. A. Rakov, and M. A. Uman
- Performance of MOV arresters during very close, direct lightning strikes to a power distribution system, IEEE Trans. on Power Delivery, Vol. 14, No. 2, April 1999, 411–18, M. I. Fernandez, K. J. Rambo, V. A. Rakov and M. A. Uman
- Characterization of the initial stage of negative rocket-triggered lightning, J. Geophys. Res., 104, 4213–22, 1999, D. Wang, V. A. Rakov, M. A. Uman, M. I. Fernandez, K. J. Rambo, G. H. Schnetzer, and R. J. Fisher

- Attachment process in rocket-triggered lightning strokes, J. Geophys. Res., 104, 2141–50, 1999, D. Wang, V. A. Rakov, M. A. Uman, N. Takagi, T. Watanabe, D. Crawford, K. J. Rambo, G. H. Schnetzer, R. J. Fisher, and Z.-I. Kawasaki

1998

- New insights into lightning processes gained from triggered-lightning experiments in Florida and Alabama, J. Geophys. Res., 103, 14,117–30 (1998), V. A. Rakov, M. A. Uman, K. J. Rambo, M. I. Fernandez, R. J. Fisher, G. H. Schnetzer, R. Thottappillil, A. Eybert-Berard, J. P. Berlandis, P. Lalande, A. Bonamy, P. Laroche, and A. Bondiou-Clergerie
- Leader properties determined with triggered lightning techniques, J. Geophys. Res., 103, 14,109–15 (1998), P. Lalande, A. Bondiou-Clergerie, P. Laroche, A. Eybert-Berard, J.-P. Berlandis, B. Bador, A. Bonamy, M. A. Uman, and V. A. Rakov
- Review and evaluation of lightning return stroke models including some aspects of their application, IEEE Trans. on EMC, vol. 40, No. 4, November 1998, part II, Special Issue on Lightning, 403–26, V. A. Rakov and M. A. Uman

1997

- Triggered-lightning experiments at Camp Blanding, Florida (1993–1995), Trans. of IEE Japan, Special Issue on Artificial Rocket Triggered Lightning, Vol. 117-B, No. 4, 446–52, 1997, M. A. Uman, V. A. Rakov, K. J. Rambo, T. W. Vaught, M. I. Fernandez, D. J. Cordier, R. M. Chandler, R. Bernstein, and C. Golden

1996

- Lightning effects studied: The underground cable program, Transmission and Distribution World, 24–33, 1996, P. P. Barker and T. A. Short
- Induced voltage measurements on an experimental distribution line during nearby rocket triggered lightning flashes, IEEE Trans. Pow. Del. 11: 980–95, 1996, P. P. Barker, T. A. Short, A. R. Eybert-Berard, and J. P. Berlandis

1995

- Mechanism of the lightning M component, J. Geophys. Res., 100, 25,701–10, 1995, V. A. Rakov, R. Thottappillil, M. A. Uman, and P. P. Barker
- Review of recent lightning research at the University of Florida, Elektrotechnik und Informationstechnik (Austria), Vol. 112, No. 6, 262–5, 1995, V. A. Rakov, M. A. Uman, and R. Thottappillil

Glossary

Atmospheric electricity sign convention: Electric field sign convention according to which a downward-directed field or field change vector is defined as positive.

Bipolar lightning: Lightning discharges sequentially transferring both positive and negative charges to the ground during the same flash.

Bremsstrahlung: Braking radiation that is emitted when a free electron is deflected in the electric field of a nucleus or, to a lesser extent, in the field of an atomic electron.

Cloud charge structure: Typically includes a net positive charge near the top, a net negative charge below it, and an additional positive charge at the bottom of the cloud.

Cloud lightning: Lightning discharges that do not involve the ground.

Cloud particles: Hydrometeors that have a fall speed lower than 0.3 m s^{-1} and remain essentially suspended or move upward in updrafts.

Compact Intracloud Discharge (CID): A small-spatial-scale (typically hundreds of meters) lightning discharge in the cloud that is thought to be the most intense natural producer of HF–VHF (3 – 300 MHz) radiation on Earth.

Continuing current: A steady current immediately following some return-stroke current pulses.

Convection mechanism: Cloud electrification mechanism in which the electric charges are supplied by external sources: fair-weather electric field and corona near the ground and cosmic rays. Organized convection provides the large-scale separation of charged particles.

Corona: A type of discharge that is confined to the immediate vicinity of an "electrode" such as a grounded object, a leader tip, the lateral surface of the leader channel, or a hydrometeor; that is, it is not a self-propagating discharge.

Cosmic rays: High-energy electromagnetic radiation and particles (mostly nuclei of hydrogen and helium) coming from outer space.

Cumulonimbus: The same as the thundercloud.

Cumulus: A type of cloud that has "puffy" or "cotton-like" appearance and is often a precursor of cumulonimbus.

Downward cloud-to-ground lightning: Lightning discharges to ground initiated by descending leaders from the cloud.

Global electric circuit: A concept according to which all the stormy weather regions worldwide (about 10 percent of the Earth's surface) constitute the global thunderstorm generator, while the fair-weather regions (about 90 percent of the globe) can be viewed as a resistive load. Lateral currents are assumed to flow freely along the highly conducting

Earth's surface and in the electrosphere (conducting region of the atmosphere just above 60 km or so).

Graupel: Millimeter-size soft hail.

Graupel-ice collision mechanism: Cloud electrification mechanism in which the electric charges are produced by collisions between precipitation particles (graupel) and cloud particles (small ice crystals) in the presence of water droplets; the large-scale separation of charged particles is provided by the action of gravity.

Hydrometeors: Various liquid and frozen water particles in the atmosphere.

Leader: Lightning process that creates a conducting path between the cloud charge source region and the ground (in the case of downward cloud-to-ground lightning) and distributes charge from the cloud source region along this path.

Lightning or lightning flash: It can be defined as a transient, high-current (typically tens of kiloamperes) electric discharge in air whose length is measured in kilometers.

M-components: Transient processes occurring in a lightning channel while it carries continuing current.

Negative lightning: Lightning discharges that effectively lower negative charge from the cloud to the ground.

Physics sign convention: Electric field sign convention according to which a downward-directed field or field change vector is defined as negative.

Positive lightning: Lightning discharges that effectively lower positive charge from the cloud to the ground.

Precipitation particles: Hydrometeors that have an appreciable fall speed (≥ 0.3 m s^{-1}). They are generally larger than cloud particles.

Return stroke: Lightning process that traverses the previously created leader channel, moving from the ground toward the cloud charge source region, and neutralizes the leader charge.

Rocket-triggered lightning: Lightning discharges artificially initiated from natural thunderclouds using the rocket-and-wire technique.

Runaway electrons: Electrons that occur when the energy gained by the free electrons between collisions, as they are accelerated by a high electric field, exceeds the energy that is lost by collisions with air molecules.

Supercell: A long-lived thunderstorm cell with a rotating updraft.

Terrestrial Gamma-ray Flashes (TGFs): Brief (typically less than 1 ms in duration) emissions of gamma-rays from thunderstorms, usually observed from space, but occasionally seen from aircraft or on the ground.

Thunderstorm cell: A unit of convection, typically some kilometers in diameter, characterized by relatively strong updrafts (≥ 10 m s^{-1}). The lifetime of an ordinary cell is of the order of 1 hour.

Transient luminous events (TLEs): Optical phenomena in the clear air between the cloud tops (at altitudes of nearly 20 km or less) and the lower ionosphere (around 60–90 km depending on the time of day). Six general types have been observed: red sprites, halos, blue starters, blue jets, gigantic jets, and elves.

Upward cloud-to-ground lightning: Lightning discharges to the ground initiated by ascending leaders from grounded objects.

References

Abarca, S. F., Corbosiero, K. L. and Galarneau, T. J. Jr (2010). An evaluation of the Worldwide Lightning Location Network (WWLLN) using the National Lightning Detection Network (NLDN) as ground truth. *J. Geophys. Res.*, 115: D18206, doi:10.1029/2009JD013411.

Ahmad, M. R., Esa, M. R. M., Cooray, V., Baharudin, Z. A. and Hettiarachchi, P. Latitude dependence of narrow bipolar pulse emissions. *Journal of Atmospheric and Solar-Terrestrial Physics*, http://dx.doi.org/10.1016/j.jastp.2015.03.005.

Anderson, R. B. and Eriksson, A. J. (1980). Lightning parameters for engineering application. *Electra*, 69: 65–102.

Aoki, M., Baba, Y. and Rakov, V. A. (2015). FDTD simulation of LEMP propagation over lossy ground: Influence of distance, ground conductivity, and source parameters. *J. Geophys. Res. Atmos.*, 120, doi:10.1002/2015JD023245.

Arabshahi, S., Dwyer, J. R., Nag, A., Rakov, V. A. and Rassoul, H. K. (2014). Numerical simulations of compact intracloud discharges as the Relativistic Runaway Electron Avalanche-Extensive Air Shower process. *J. Geophys. Res.*, 119(1): 479–89, doi:10.1002/2013JA018974.

Avila, E. E., Burgesser, R. E., Castellano, N. E., Pereyra, R. G. and Saunders, C. P. (2011). Charge separation in low-temperature ice cloud regions. *J. Geophys. Res.*, 116: D14202, doi:10.1029/2010JD015475.

Baba, Y. and Rakov, V. A. (2005a). On the use of lumped sources in lightning return stroke models. *J. Geophys. Res.*, 110: D03101, doi:10.1029/2004JD005202.

Baba, Y and Rakov, V. A. (2005b). On the mechanism of attenuation of current waves propagating along a vertical perfectly conducting wire above ground: application to lightning. *IEEE Trans. on EMC*, 47(3): 521–32.

Baba, Y. and Rakov, V. A. (2007a). Electromagnetic fields at the top of a tall building associated with nearby lightning return strokes. *IEEE Trans. Electromagn. Compat.*, 49(3): 632–43.

Baba, Y. and Rakov, V. A. (2007b). Influences of the presence of a tall grounded strike object and an upward connecting leader on lightning currents and electromagnetic fields. *IEEE Trans. on EMC*, 49(4): 886–92.

Baharudin, Z. A., Ahmad, N. A., Makela, J. S., Fernando, M. and Cooray, V. (2014). Negative cloud-to-ground lightning flashes in Malaysia. *J. Atmos. Solar-Terrestr. Phys.*, 108: 61–7.

Baker, M. B. and Dash, J. G. (1994). Mechanism of charge transfer between colliding ice particles in thunderstorms. *J. Geophys. Res.*, 99: 10,621–6.

Ballarotti, M. G., Medeiros, C., Saba, M. M. F., Schulz, W. and Pinto, O. Jr (2012). Frequency distributions of some parameters of negative downward lightning flashes based on accurate-stroke-count studies. *J. Geophys. Res.*, 117: D06112, doi:10.1029/2011JD017135.

Bazelyan, E. M. and Raizer, Y. P. (2000). *Lightning Physics and Lightning Protection*. Bristol: Institute of Physics Publication, 325 pp.

Beasley, W. H., Uman, M. A. and Rustan, P. L. (1982). Electric fields preceding cloud to ground lightning flashes. *J. Geophys. Res.*, 87: 4884–902.

Berger, G., Hermant, A. and Labbe, A. S. (1996). Observations of natural lightning in France. In *Proc. of the 23rd Int. Conf. on Lightning Protection*, vol. 1, Florence, Italy, 67–72.

Berger, K. (1972). Mesungen und Resultate der Blitzforschung auf dem Monte San Salvatore bei Lugano. der Jahre 1963–1971. *Bull. SEV*, 63: 1403–22.

Berger, K. and Vogelsanger, E. (1966). Photographische Blitzuntersuchungen der Jahre 1955–1965 auf dem Monte San Salvatore. *Bull. Schweiz. Elektrotech. Ver.*, 57: 599–620.

Berger, K., Anderson, R. B. and Kroninger, H. (1975). Parameters of lightning flashes. *Electra*, 80: 223–37.

Betz, H.-D., Schmidt, K. and Oettinger, W. P. (2009). LINET—an international VLF/LF lightning detection network in Europe. In Betz, H.-D., Schumann, U. and Laroche, P. (eds), *Lightning: Principles, Instruments and Applications*, Ch. 5, Dordrecht, NL: Springer.

Bevington, P. R. (1969). *Data Reduction and Error Analysis for the Physical Sciences*. New York: McGraw-Hill.

Biagi, C. J., Cummins, K. L., Krider, E. P. and Kehoe, K. E. (2007). NLDN performance in Southern Arizona, Texas and Oklahoma in 2003–2004. *J. Geophys. Res.*, 112: D05208. doi:1029/2006JD007341.

Biagi, C. J., Jordan, D. M., Uman, M. A., Hill, J. D., Beasley, W. H. and Howard, J. (2009). High-speed video observations of rocket-and-wire initiated lightning. *Geophys. Res. Lett.*, 36: L15801, doi:10.1029/2009GL038525.

Biagi, C. J., Uman, M. A., Hill, J. D., Jordan, D. M., Rakov, V. A. and Dwyer, J. R. (2010). Observations of stepping mechanisms in a rocket-and-wire triggered lightning flash. *J. Geophys. Res.*, 115: D23215. doi:10.1029/2010JD014616.

Brook, M. (1992). Breakdown electric fields in winter storms. *Res. Lett. Atmos. Electr.*, 12: 47–52.

Bruce, C. E. R. and Golde, R. H. (1941). The lightning discharge. *J. Inst. Elec. Eng.*, 88: 487–520.

Chauzy, S. and Soula, S. (1999). Contribution of the ground corona ions to the convective charging mechanism. *Atmos. Res.*, 51: 279–300.

Chen, Y., Wang, X. and Rakov, V. A. (2015). Approximate expressions for lightning electromagnetic fields at near and far ranges: Influence of return-stroke speed. *J. Geophys. Res. Atmos.*, 120: 2855–80. doi:10.1002/2014JD022867.

Cianos, N., Oetzel, G. N. and Pierce, E. T. (1972). A technique for accurately locating lightning at close ranges. *J. Appl. Meteor.*, 11: 1120–7.

CIGRE WG 33.01, Report 63. 1991. Guide to procedures for estimating the lightning performance of transmission lines, 61 pp.

CIGRE WG C4.407 Technical Brochure 549, lightning parameters for engineering applications, V. A. Rakov, Convenor (US), A. Borghetti, Secretary (IT), C. Bouquegneau (BE), W. A. Chisholm (CA), V. Cooray (SE), K. Cummins (US), G. Diendorfer (AT), F. Heidler (DE), A. Hussein (CA), M. Ishii (JP), C. A. Nucci (IT), A. Piantini (BR), O. Pinto, Jr (BR), X. Qie (CN), F. Rachidi (CH), M. M. F. Saba (BR), T. Shindo (JP), W. Schulz (AT), R. Thottappillil (SE), S. Visacro (BR), W. Zischank (DE), 117 pp., August 2013.

Clarence, N. D. and Malan, D. J. (1957). Preliminary discharge processes in lightning flashes to ground. *Q. J. Roy. Meteorol. Soc.*, 83: 161–72.

Cooray, V., Cooray, G., Marshall, T., Arabshahi, S., Dwyer, J. and Rassoul, H. (2014). Electromagnetic fields of a relativistic electron avalanche with special attention to the origin of lightning signatures known as narrow bipolar pulses. *Atmos. Res.*, 149: 346–58.

Cooray, V. and Jayaratne, K. P. S. C. (1994). Characteristics of lightning flashes observed in Sri Lanka in the tropics. *J. Geophys. Res.*, 99: 21,051–6.

Cooray, V., Montano, R. and Rakov, V. (2004). A model to represent negative and positive lightning first return strokes with connecting leaders. *J. Electrostatics*, 60: 97–109.

Cooray, V. and Perez, H. (1994b). Some features of lightning flashes observed in Sweden. *J. Geophys. Res.*, 99: 10,683–8.

Crawford, D. E., Rakov, V. A., Uman, M. A., Schnetzer, G. H., Rambo, K. J., Stapleton, M. V. and Fisher, R. J. (2001). The close lightning electromagnetic environment: Dart-leader electric field change versus distance. *J. Geophys. Res.*, 106: 14,909–17.

Cummins, K. L. and Murphy, M.J. (2009). An overview of lightning locating systems: history, techniques, and data uses, with an in-depth look at the U.S. NLDN. *IEEE Trans. on EMC*, 51(3): 499–518.

da Silva, C. L. and Pasko, V. P. (2015). Physical mechanism of initial breakdown pulses and narrow bipolar events in lightning discharges. *J. Geophys. Res. Atmos.*, 120: 4989–5009, doi:10.1002/2015JD023209.

Davis, R. (1962). Frequency of lightning flashover on overhead lines. In *Gas Discharges and the Electricity Supply Industry*. London: Butterworth, 125–38.

Demetriades, N. W. S., Murphy, M. J. and Cramer, J. A. (2010). Validation of Vaisala's Global Lightning Dataset (GLD360) over the continental United States. *Preprints, 29th Conference Hurricanes and Tropical Meteorology, 10–14 May,* Tucson, AZ, 6 pp.

Diendorfer, G., Pichler, H. and Mair, M. (2009). Some parameters of negative upward-initiated lightning to the Gaisberg tower (2000–2007). *IEEE Trans. Electromagn. Compat.*, 51: 443–52.

Diendorfer, G. and Uman, M. A. (1990). An improved return stroke model with specified channel-base current. *J. Geophys. Res.*, 95: 13,621–44.

Dowden, R. L., Brundell, J. B. and Rodger, C. J. (2002). VLF lightning location by time of group arrival (TOGA) at multiple sites. *J. Atmos. Solar Terr. Phys.*, 64(7): 817–30.

Dwyer, J. R. and Babich, L. P. (2011). Low-energy electron production by relativistic runaway electron avalanches in air. *J. Geophys. Res.*, 116: A09301, doi:10.1029/2011JA016494.

Dwyer, J. R. and Babich, L. (2012). Reply to comment by A. V. Gurevich et al. on "Low-energy electron production by relativistic runaway electron avalanches in air." *J. Geophys. Res.*, 117: A04303, doi:10.1029/2011JA017487.

Dwyer, J. R. and Cummer, S. A. (2013). Radio emissions from terrestrial gamma-ray flashes. *J. Geophys. Res. Space Physics*, 118: 3769–90, doi:10.1002/jgra.50188.

Eack, K. B. (2004). Electrical characteristics of narrow bipolar events. *Geophys. Res. Lett.*, 31: L20102, doi:10.1029/2004GL021117.

Eidelman, S. et al. (2004). Review of Particle Physics. *Physics Letters B*, 592: 1–1109, doi:10.1016/j.physletb.2004.0 6.001.

Eriksson, A. J. (1978). Lightning and tall structures. *Trans. S. Afr. Inst. Electr. Eng.*, 69(8): 238–52.

Eriksson, A. J. (1987). The incidence of lightning strikes to power lines. *IEEE Trans. Power Deliv.*, 2(3): 859–70.

Fisher, R. J., Schnetzer, G. H. and Morris, M. E. (1994). Measured fields and earth potentials at 10 and 20 meters from the base of triggered-lightning channels. In *Proc. 22nd Int. Conf. on Lightning Protection*. Budapest, Hungary, Paper R 1c-10, 6 p.

Fisher, R. J., Schnetzer, G. H., Thottappillil, R., Rakov, V. A., Uman, M. A. and Goldberg, J. D. (1993). Parameters of triggered-lightning flashes in Florida and Alabama. *J. Geophys. Res.*, 98: 22,887–902.

Fleenor, S. A., Biagi, C. J., Cummins, K. L., Krider, E. P., Shao, X.-M. (2009). Characteristics of cloud-to-ground lightning in warm-season thunderstorms in the Great Plains. *Atmos Res.*, 91: 333–52.

Garbagnati, E., Guidice, E., Lo Piparo, G. B. and Magagnoli, U. (1974). Survey of the characteristics of lightning stroke currents in Italy – Results obtained in the years from 1970 to 1973. *ENEL Report* R5/63–27.

Golde, R. H. (1945). The frequency of occurrence and the distribution of lightning flashes to transmission lines. *AIEE Trans.*, 64(3): 902–10.

Golde, R. H. (1977). The lightning conductor. Lightning. In Golde R. H. (ed.), *Lightning Protection*. New York: Academic Press, 2, 545–76.

Gorin, B. N. (1972). Lightning discharges to the Ostankino television tower. *Elecktr.* No. 2: 24–9.

Gorin, B. N., Levitov, V. I. and Shkilev, A. V. (1976). Distinguishing features of lightning strokes to high constructions. 4th Int. Conf. on Gas Discharges. *IEE Conf. Publ.*, 43, 271–3.

Gorin, B. N., Levitov, V. I. and Shkilev, A. V. (1976). Some principles of leader discharge of air gaps with a strong non-uniform field. 4th Int. Conf. on Gas Discharges. *IEE Conf. Publ.*, 143, 274–8.

Gorin, B. N. and Shkilev, A. V. (1984). Measurements of lightning currents at the Ostankino tower. *Elektrichestvo*, 8: 64–5.

Griffiths, R. F. and Phelps, C. T. (1976a). The effects of air pressure and water vapor content on the propagation of positive corona streamers, and their implications to lightning initiation. *Q. J. R. Meteor. Soc.*, 102: 419–26.

Griffiths, R. F. and Phelps, C. T. (1976b). A model of lightning initiation arising from positive corona streamer development. *J. Geophys. Res.*, 31: 3671–6.

Gunn, R. (1948). Electric field intensity inside of natural clouds. *J. Appl. Phys.*, 19: 481–4.

Gurevich, A. V. and Karashtin, A. N. (2013). Runaway breakdown and hydrometeors in lightning initiation, *Phys. Rev. Lett.*, 110(18): 185005, doi:10.1103/PhysRevLett.110.185005.

Gurevich, A. V., Milikh, G.M. and Roussel-Dupre, R. (1992). Runaway electron mechanism of air breakdown and preconditioning during a thunderstorm. *Phys. Lett. A*, 165:
463–7.

Gurevich, A. V., Roussel-Dupre, R., Zybin, K. P. and Milikh, G. M. (2012). Comment
on "Low-energy electron production by relativistic runway electron avalanches in
air" by J. R. Dwyer and L. P. Babich. *J. Geophys. Res.*, 117: A04302, doi:10.1029/
2011JA017431.

Gurevich, A. V. and Zybin, K. P. (2001). Runaway breakdown and electric discharges in
thunderstorms. *Physics – Uspekhi*, 44(11): 1119–40.

Gurevich, A. V., Zybin, K. P. and Roussel-Dupre, R. A. (1999). Lightning initiation by
simultaneous effect of runaway breakdown and cosmic ray showers. *Phys. Lett. A*, 254:
79–87.

Heckman, S. and Liu, C. (2010). The application of total lightning detection and cell
tracking for severe weather prediction. In *Proceedings of GROUND'2010 & 4th LPE*,
Salvador, Brazil, November, 234–40.

Heidler, F. (1985). Traveling current source model for LEMP calculation. In *Proc. 6th Int.
Zurich Symp. on Electromagnetic Compatibility*, Zurich, Switzerland, 157–62.

Hendry, J. (1993). Panning for lightning (including comments on the photos by
M. A. Uman). *Weatherwise*, 45(6): 19.

Hermant, A. (2000). *Traqueur d'Orages*. Paris, France: Nathan/HER.

Hill, J. D., Pilkey, J., Uman, M. A., Jordan, D. M., Rison, W. and Krehbiel, P. R. (2012).
Geometrical and electrical characteristics of the initial stage in Florida triggered lightning. *Geophys. Res. Lett.*, 39: L09807, doi:10.1029/2012GL051932.

Holle, R. L. and Lopez, R. E. (1993). Overview of real-time lightning detection systems and
their meteorological uses. *NOAA Technical Memorandum*, ERL NSSL-102, 68.

Holle, R. L., López, R. E. and Zimmerman, C. (1999). Updated recommendations for
lightning safety. *Bull. Am. Meteor. Soc.*, 80: 2035–41.

Horner, F. (1954). The accuracy of the location sources of atmospherics by radio direction
finding. *Proceedings of the IEE-Part III: Radio and Communication Engineering*, 101:
383–90.

Horner, F. (1957). Very-low-frequency propagation and direction finding. *Proceedings of
the IEE-Part B: Radio and Electronic Engineering*, 101B: 73–80.

Horvath, T. (2000). Rolling Sphere – Theory and Application, in *Proc. 25th Int. Conf. on
Lightning Protection*, Rhodes, Greece, 301–5.

Hutchins, M. L., Holzworth, R. H., Brundell, J. B. and Rodger, C. J. (2012a). Relative
detection efficiency of the World Wide Lightning Location Network. *Radio. Sci.*, 47:
RS6005, doi:10.1029/2012RS005049.

Hutchins, M. L., Holzworth, R. H., Rodger, C. J., and Brundell, J. B. (2012b). Far field
power of lightning strokes as measured by the World Wide Lightning Location Network.
J. Atmos. Oceanic. Technol., 29: 1102–10, doi:10.1175/JTECH-D-11-00174.1.

Idone, V. P. and Orville, R. E. (1982). Lightning return stroke velocities in the
Thunderstorm Research International Program (TRIP). *J. Geophys. Res.*, 87: 4903–15.

IEC 62305–1, Protection against lightning – Part 1: General principles, Ed. 2, (2010),
Geneva: International Electrotechnical Commission.

IEEE Standard 1243–1997. (1997). IEEE Guide for Improving the Lightning Performance of Transmission Lines.

IEEE Standard 1410–2010. (2010). IEEE Guide for Improving the Lightning Performance of Electric Power Overhead Distribution Lines.

Jacobson, E. A. and Krider, E. P. (1976). Electrostatic field changes produced by Florida lightning. *J. Atmos. Sci.*, 33: 113–7.

Jacobson, A. R. and Light, T. E. L. (2012). Revisiting "Narrow Bipolar Event" intracloud lightning using the FORTE satellite. *Ann. Geophys.*, 30(2): 389–404, doi:10.5194/angeo-30-389-2012.

Jacobson, A. R., Light, T. E. L., Hamlin, T. and Nemzek, R. (2013). Joint radio and optical observations of the most radio-powerful intracloud lightning discharges. *Ann. Geophys.*, 31: 563–580, doi:10.5194/angeo-31-563-2013.

Jayaratne, E. R., Saunders, C. P. R. and Hallett, J. (1983). Laboratory studies of the charging of soft-hail during ice crystal interactions. *Q. J. R. Meteor. Soc.*, 109: 609–30.

Jerauld, J., Rakov, V. A., Uman, M. A., Rambo, K. J., Jordan, D. M., Cummins, K. L., Cramer, J. A. (2005). An evaluation of the performance characteristics of the U.S. National Lightning Detection Network in Florida using rocket-triggered lightning. *J. Geophys. Res.*, 110: D19106, doi:10.1029/2005JD005924.

Jiang, R., Qie, X., Yang, J., Wang, C. and Zhao, Y. (2013). Characteristics of M-component in rocket-triggered lightning and a discussion on its mechanism. *Radio Sci.*, 48: 597–606, doi:10.1002/rds.20065.

Jordan, D. M., Rakov, V. A., Beasley, W. H. and Uman, M. A. (1997). Luminosity characteristics of dart leaders and return strokes in natural lightning. *J. Geophys. Res.*, 102: 22, 25–32.

Karunarathne, S., Marshall, T. C., Stolzenburg, M. and Karunarathna, N. (2015). Observations of positive narrow bipolar pulses. *J. Geophys. Res. Atmos.*, 120, doi:10.1002/2015JD023150.

Kidder, R. E. (1973). The location of lightning flashes at ranges less than 100 km. *J. Atmos. Terr. Phys.*, 35: 283–90.

Kitagawa, N., Brook, M. and Workman, E. J. (1962). Continuing currents in cloud-to-ground lightning discharges. *J. Geophys. Res.*, 67: 637–47.

Koshak, W. J. and Krider, E. P. (1989). Analysis of lightning field changes during active Florida thunderstorms. *J. Geophys. Res.*, 94: 1165–86.

Kostinskiy, A. Yu., Syssoev, V. S., Bogatov, N. A., Mareev, E. A., Andreev, M. G., Makalsky, L. M., Sukharevsky, D. I. and Rakov, V. A. (2015). Observation of a New Class of Electric Discharges within Artificial Clouds of Charged Water Droplets and Its Implication for Lightning Initiation within Thunderclouds. *Geophys. Res. Lett.*, 42, 8165–8171, doi: 10.1002/2015GL065620.

Krehbiel, P. R. (1986). The electrical structure of thunderstorms. In Krider, E. P. and Roble, R. G. (eds), *The Earth's Electrical Environment*, eds. E. P. Krider and R. G. Roble, 90–113. Washington, DC: National Academy Press.

Krehbiel, P. R., Brook, M. and McCrory, R. A. (1979). An analysis of the charge structure of lightning discharges to the ground. *J. Geophys. Res.*, 84: 2432–56.

Krehbiel, P. R., Riousset, J. A., Pasko, V. P., Thomas, R. J., Rison, W., Stanley, M. A. and Edens, H. E. (2008). Upward electrical discharges from thunderstorms. *Nature Geosci.*, 1: 233–37, doi:10.1038/ngeo162.

Krider, E. P. (1994). On the peak electromagnetic fields radiated by lightning return strokes toward the middle-atmosphere. *J. Atmos. Electr.*, 14: 17–24.

Krider, E. P., Noggle, R. C., and Uman, M. A. (1976). A gated wideband magnetic direction finder for lightning return strokes. *J. Appl. Meteor.*, 15: 301–6.

Krider, E. P., Weidman, C. D. and Noggle, R. C. (1977). The electric field produced by lightning stepped leaders. *J. Geophys. Res.*, 82: 951–60.

Kuzhekin, I. P., Larionov, V. P., and Prokhorov, E. N. (2003). *Lightning and Lightning Protection*. Moscow: Znak, 300.

Lebedev, V. B., Feldman, G. G., Gorin, B. N., Shcherbakov, Yu. V., Syssoev, V. S., Rakov, V. A., Uman, M. A. and Olsen, R. C. (2007). Test of the image converter camera complex for research of discharges in long air gaps and lightning. In *Proceedings of the 13th international conference on atmospheric electricity*, Beijing, China, August 13–17, 509–12.

Lee, R. H. (1978). Protection zone for buildings against lightning strokes using transmission line protection practice. *IEEE Trans. Ind. Appl.*, 14: 465–70.

Lennon, C. L. and Poehler, H. A. (1982). Lightning detection and ranging. *Astronaut Aeronautics*, 20: 29–31.

Leteinturier, C., Hamelin, J. H. and Eybert-Berard, A. (1991). Submicrosecond characteristics of lightning return-stroke currents. *IEEE Trans. Electromagn. Compat.*, 33: 351–7.

Lewis, E. A., Harvey, R. B. and Rasmussen, J. E. (1960). Hyperbolic direction finding with sferics of transatlantic origin. *J. Geophys. Res.*, 65: 1879–1905.

Lin, Y. T., Uman, M. A., Tiller, J. A., Brantley, R. D., Beasley, W. H., Krider, E. P. and Weidman, C. D. (1979). Characterization of lightning return stroke electric and magnetic fields from simultaneous two-station measurements. *J. Geophys. Res.*, 84: 6307–14.

Liu, H., Dong, W., Wu, T., Zheng, D. and Zhang, Y. (2012). Observation of compact intracloud discharges using VHF broadband interferometers. *J. Geophys. Res.*, 117: D01203, doi:10.1029/2011JD016185.

Loeb, L. B. (1966). The mechanisms of stepped and dart leaders in cloud-to-ground lightning strokes. *J. Geophys. Res.*, 71: 4711–21.

Lojou, J.-Y., Murphy, M. J., Holle, R. L. and Demetriades, N. W. S. (2009). Nowcasting of thunderstorms using VHF measurements. In Betz, H.-D., Schumann, U. and Laroche, P. (eds), *Lightning: Principles, Instruments and Applications*, Ch. 11, Dordrecht, NL: Springer.

Love, E. R. (1973). Improvements on lightning stroke modeling and applications to the design of EHV and UHV transmission lines. M.Sc. Thesis, University of Colorado.

Lu, G. (2006). Transient electric field at high altitudes due to lightning: Possible role of induction field in the formation of elves. *J. Geophys. Res.*, 111: D02103, doi:10.1029/2005JD005781.

Lu, G. et al. (2011). Lightning development associated with two negative gigantic jets. *Geophys. Res. Lett.*, 38: L12801, doi:10.1029/2011GL047662.

Lu, W., Chen, L., Ma, Y., Rakov, V. A., Gao, Y., Zhang, Y., Yin, Q. and Zhang, Y. (2013). Lightning attachment process involving connection of downward negative leader to the lateral surface of upward connecting leader. *Geophys. Res. Lett.*, 40: 5531–35, doi:10.1002/2013GL058060

Lü, F., Zhu, B., Zhou, H., Rakov, V. A., Xu, W. and Qin, Z. (2013). Observations of compact intracloud lightning discharges in northernmost region (51° N) of China. *J. Geophys. Res. Atmos.*, 118: 4458–65, doi:10.1002/jgrd.50295.

MacGorman, D. R., Maier, M. W. and Rust, W. D. (1984). *Lightning strike density for the contiguous United States from thunderstorm duration records,* NUREG/CR-3759, Office of Nuclear Regulatory Research, U.S. Nuclear Regulatory Commission, 44 p., Washington, DC.

MacGorman, D. R. and Rust, W. D. (1998). *The Electrical Nature of Thunderstorms*, 422 p., New York: Oxford University Press.

Mach, D. M., MacGorman, D. R., Rust, W. D. and Arnold, R. T. (1986), Site errors and detection efficiency in a magnetic direction-finder network for locating lightning strikes to ground. *J. Atmos. Ocean. Tech.*, 3: 67–74.

Malan, D. J. (1952). Les déchargesdansl'airet la charge inférieure positive d'un nuageorageuse. *Ann. Geophys.*, 8: 385–401.

Mallick, S. et al. (2014a). Performance characteristics of the NLDN for return strokes and pulses superimposed on steady currents, based on rocket-triggered lightning data acquired in Florida in 2004–2012. *J. Geophys. Res. Atmos.*, 119: 3825–56, doi:10.1002/2013JD021401.

Mallick, S., Rakov, V. A., Hill, J. D., Ngin, T., Gamerota, W. R., Pilkey, J. T., Jordan, D. M., Uman, M.A., Heckman, S., Sloop, C. D. and Liu, C. (2015). Performance characteristics of the ENTLN evaluated using rocket-triggered lightning data. *Electric Power Systems Research*, 118: 15–28.

Mallick, S., Rakov, V. A., Ngin, T., Gamerota, W. R., Pilkey, J. T., Hill, J. D., Uman, M. A., Jordan, D. M., Cramer, J. A. and Nag, A. (2014a). An Update on the Performance Characteristics of the NLDN, *2014 ILDC/ILMC*, Tucson, Arizona, USA.

Mallick, S., Rakov, V. A., Ngin, T., Gamerota, W. R., Pilkey, J. T., Hill, J. D., Uman, M. A., Jordan, D. M., Nag, A. and Said, R. K. (2014b). Evaluation of the GLD360 performance characteristics using rocket-and-wire triggered lightning data. *Geophys. Res. Lett.*, 41: 3636–42, doi:10.1002/2014GL059920.

Mallick, S., Rakov, V. A., Ngin, T., Gamerota, W. R., Pilkey, J. T., Hill, J. D., Uman, M. A., Jordan, D. M., Hutchins, M. L. and Holzworth, R. H. (2014). Evaluation of the WWLLN Performance Characteristics Using Rocket-triggered Lightning Data, *2014 GROUND/LPE*, Manaus, Brazil.

Mallick, S., Rakov, V. A., Tsalikis, D., Nag, A., Biagi, C., Hill, D., Jordan, D. M., Uman, M. A. and Cramer, J. A. (2014). On remote measurements of lightning return stroke peak currents. *Atmos. Res.*, 135–136: 306–13.

Marshall, R. A., da Silva, C. L. and Pasko, V. P. (2015). Elve doublets and compact intracloud discharges. *Geophys. Res. Lett.*, 42: 6112–19, doi:10.1002/2015GL064862.

Marshall, T. C. and Rust, W. D. (1991). Electric field soundings through thunderstorms. *J. Geophys. Res.*, 96: 22,297–306.

Maslowski, G. and Rakov, V. A. (2007). Equivalency of lightning return stroke models employing lumped and distributed current sources. *IEEE Trans. on EMC*, 49(1), February 2007: 123–32.

Mason, B. L. and Dash, J. G. (2000). Charge and mass transfer in ice-ice collisions: Experimental observations of a mechanism in thunderstorm electrification. *J. Geophys. Res.*, 105: 20,185–92.

Mazur, V. and Ruhnke, L. (1993). Common physical processes in natural and artificially triggered lightning. *J. Geophys. Res.*, 98: 12,913–30.

McCann, E. D. (1944). The measurement of lightning currents in direct strokes. *AIEE Trans.*, 63: 1157–64.

Miki, M., Rakov, V. A., Rambo, K. J., Schnetzer, G. H. and Uman, M. A. (2002). Electric fields near triggered lightning channels measured with Pockels sensors. *J. Geophys. Res.*, 107(D16): 10.1029/2001JD001087, 11 p.

Moore, C. B. and Vonnegut, B. (1977). The thundercloud. In R. H. Golde (ed.), *Lightning, Vol. 1, Physics of Lightning*, 51–98. New York: Academic Press.

Moore, C. B., Vonnegut, B. and Holden, D. N. (1989). Anomalous electric fields associated with clouds growing over a source of negative space charge. *J. Geophys. Res.*, 94: 13,127–34.

Morimoto, T., Hirata, A., Kawasaki, Z., Ushio, T., Matsumoto, A. and Lee, J. H. (2004). An operational VHF broadband digital interferometer for lightning monitoring. *IEEJ Trans. Fundam. Mater.* 124(12): 1232–8.

Murphy, M. and Nag, A. (2015). Cloud lightning performance and climatology of the U.S. based on the upgraded U.S. National Lightning Detection Network, in *Seventh Conference on the Meteorological Applications of Lightning Data*, paper 8.4, Phoenix, AZ: Am. Meteorol. Soc., 4–8 January.

Naccarato, K. P., Pinto, O. Jr, Garcia, S. A. M., Murphy, M., Demetriades, N. and Cramer, J. (2010). *Validation of the New GLD360 Dataset in Brazil: First Results, ILDC.* Orlando, FL, 19–22 July, 6 p.

Nag, A. and Rakov, V. A. (2009). Some inferences on the role of lower positive charge region in facilitating different types of lightning. *Geophys. Res. Lett.*, 36: L05815, doi:10.1029/2008GL036783.

Nag, A. and Rakov, V. A. (2010). Compact intracloud lightning discharges: 1. Mechanism of electromagnetic radiation and modeling. *J. Geophys. Res.*, 115: D20102, doi:10.1029/2010JD014235.

Nag, A. and Rakov V. A. (2012). Positive lightning: An overview, new observations, and inferences. *J. Geophys. Res.*, 117: D08109, doi:10.1029/2012JD017545.

Nag, A., Rakov, V. A. and Cramer, J. A. (2011). Remote measurements of currents in cloud lightning discharges. *IEEE Trans. Electromagn. Compat.*, 53(2): 407–13.

Nag, A., Murphy, M. J., Cummins, K. L., Pifer, A. E. and Cramer, J. A. (2014). Recent evolution of the U.S. National Lightning Detection Network, in *23rd International Lightning Detection Conference & 5th International Lightning Meteorology Conference*. Tucson, AZ: Vaisala Inc..

Nag, A., Murphy, M. J., Schulz, W. and Cummins, K. L. (2015). Lightning locating systems: Insights on characteristics and validation techniques. *Earth and Space Science*, 2: 65–93, doi:10.1002/2014EA000051.

Nag, A., Rakov, V. A., Tsalikis, D. and Cramer, J. A. (2010). On phenomenology of compact intracloud lightning discharges. *J. Geophys. Res.*, 115, D14115: doi:10.1029/2009JD012957.

Nag, A. et al. (2011). Evaluation of U.S. National Lightning Detection Network performance characteristics using rocket-triggered lightning data acquired in 2004–2009. *J. Geophys. Res.*, 116: D02123, doi:10.1029/2010JD014929.

NFPA 780, National Fire Protection Association, Standard for the installation of lightning protection systems, 2014 edition, Available from NFPA, 1 Battery march Park, PO Box 9101, Quincy, Massachusetts 02169–7471, USA.

Nguyen, M. D. and Michnowski, S. (1996). On the initiation of lightning discharge in a cloud 2. The lightning initiation on precipitation particles. *J. Geophys. Res.*, 101: 26,675–80.

Nishino, M., Iwai, A. and Kashiwagi, M. (1973). Location of the sources of atmospherics in and around Japan. In *Proceedings of Research Institute Atmospherics*. Japan: Nagoya University, 20: 9–21.

Nucci, C. A., Diendorfer, G., Uman, M. A., Rachidi, F., Ianoz, M. and Mazzetti, C. (1990). Lightning return stroke current models with specified channel-base current: A review and comparison. *J. Geophys. Res.*, 95: 20395–408.

Nucci, C. A., Mazzetti, C., Rachidi, F. and Ianoz, M. (1988). On lightning return stroke models for LEMP calculations. In *Proc. 19th Int. Conf. on Lightning Protection*, Graz, Austria, 463–9.

Olsen, R. C., Jordan, D. M., Rakov, V. A., Uman, M. A. and Grimes, N. (2004). Observed two-dimensional return stroke propagation speeds in the bottom 170 m of a rocket-triggered lightning channel. *Geophys. Res. Lett.*, 31: L16107, doi:10.1029/2004GL020187.

Orville, R. E. (2008). Development of the National Lightning Detection Network. *Bull Am. Meteorol. Soc.*, 89(2): 180–90.

Phelps, C. T. (1974). Positive streamers system intensification and its possible role in lightning initiation. *J. Atmos. Terr. Phys.*, 36: 103–11.

Pierce, E. T. (1972). Triggered lightning and some unsuspected lightning hazards. *Naval Res. Rev.*, 25: 14–28.

Pierce, E. T. (1974). Atmospheric electricity – some themes. *Bull. Am. Meteor. Soc.*, 55: 1186–94.

Pierce, E. T. (1977). Atmospherics and radio noise. In R. H. Golde (ed.), *Lightning*, Vol. 1, Physics of Lightning. New York, NY: Academic Press, 351–84.

Poelman, D. R., Schulz, W. and Vergeiner, C. (2013). Performance characteristics of distinct lightning detection networks covering Belgium. *J. Atmos. Ocean. Technol.*, 30: 942–51, doi:10.1175/JTECH-D-12-00162.1.

Ponjola, H., and Makela, A. (2013). The comparison of GLD360 and EUCLID lightning location systems in Europe. *Atmos. Res.*, 123: 117–28.

Proctor, D. E. (1971). A hyperbolic system for obtaining VHF radio pictures of lightning. *J. Geophys. Res.*, 76: 1478–89.

Qie, X., Yu, Y., Wang, D., Wang, H. and Chu, R. (2002). Characteristics of Cloud-to-Ground Lightning in Chinese Inland Plateau. *J. Meteorol. Soc. Jpn,* 80: 745–54.

Rachidi, F., Rakov, V. A., Nucci, C. A. and Bermudez, J. L. (2002). The Effect of Vertically-Extended Strike Object on the Distribution of Current Along the Lightning Channel. *J. Geophys. Res.*, 107(D23): 4699, doi:10.1029/2002JD002119.

Rakov, V. A. (1985). On estimating the lightning peak current distribution parameters taking into account the lower measurement limit. *Elektrichestvo*, 2: 57–9.

Rakov, V. A. (1997). Lightning electromagnetic fields: Modeling and measurements. In *Proc. 12th Int. Zurich Symp. on Electromagnetic Compatibility*, Zurich, Switzerland, 59–64.

Rakov, V. A. (2003). A review of the interaction of lightning with tall objects. *Recent Res. Devel. Geophysics*, 5: 57–71, Research Signpost, India.

Rakov, V. A. (2007). Lightning Return Stroke Speed. *J. Lightning Research*, 1: 80–9.

Rakov, V. A. (2013). Electromagnetic methods of lightning detection. *Surveys in Geophysics*, 34(6): 731–53, DOI 10.1007/s10712–013–9251–1.

Rakov, V. A. and Dulzon, A. A. (1987). Calculated electromagnetic fields of lightning return stroke. *Tekh. Elektrodinam.*, 1: 87–9.

Rakov, V. A. and Dulzon, A. A. (1991). A modified transmission line model for lightning return stroke field calculations. In *Proc. 9th Int. Zurich. Symp. on Electromagnetic Compatibility*, Zurich, Switzerland, 229–35.

Rakov, V. A., Thottappillil, R. and Uman, M. A. (1992). On the empirical formula of Willett et al. relating lightning return-stroke peak current and peak electric field. *J. Geophys. Res.*, 97: 11,527–33.

Rakov, V. A., Thottappillil, R., Uman, M. A. and Barker, P. P. (1995). Mechanism of the lightning M component. *J. Geophys. Res.*, 100: 25,701–10.

Rakov, V. A. and Tuni, W. G. (2003). Lightning electric field intensity at high altitudes: Inferences for production of Elves. *J. Geophys. Res.*, 108(D20): 4639, doi:10.1029/2003JD003618.

Rakov, V. A. and Uman, M. A. (1990a). Long continuing current in negative lightning ground flashes. *J. Geophys. Res.*, 95: 5455–70.

Rakov, V. A. and Uman, M. A. (1990b). Some properties of negative cloud-to-ground lightning flashes versus stroke order. *J. Geophys. Res.*, 95: 5447–53.

Rakov, V. A. and Uman, M. A. (1990c). Waveforms of first and subsequent leaders in negative lightning flashes. *J. Geophys. Res.*, 95: 16,561–77.

Rakov, V. A. and Uman, M. A. (2003). *Lightning: Physics and Effects*, 687 p., New York, NY: Cambridge.

Rakov, V. A., Uman, M. A., Jordan, D. M. and Priore, C. A. III (1990). Ratio of leader to return stroke field change for first and subsequent lightning strokes. *J. Geophys. Res.*, 95: 16,579–87.

Rakov, V. A., Uman, M. A., Rambo, K. J., Fernandez, M. I., Fisher, R. J., Schnetzer, G. H., Thottappillil, R., Eybert-Berard, A., Berlandis, J. P., Lalande, P., Bonamy, A., Laroche, P. and Bondiou-Clergerie, A. (1998). New insights into lightning processes gained from triggered-lightning experiments in Florida and Alabama. *J. Geophys. Res.*, 103: 14,117–30.

Ray, P. S., MacGorman, D. R., Rust, W. D., Taylor, W. L. and Rasmussen, L. W. (1987). Lightning location relative to storm structure in a super cell storm and a multi cell storm. *J. Geophys. Res.*, 92: 5713–24.

Rison, W., Thomas, R. J., Krehbiel, P. R., Hamlin, T. and Harlin, J. (1999). A GPS-based three-dimensional lightning mapping system: initial observations in central New Mexico. *Geophys. Res. Lett.*, 26: 3573–6.

Romps, D. M., Seeley, J. T., Vollaro, D. and Molinari, J. (2014). Projected increase in lightning strikes in the United States due to global warming. *Science*, 346(6211): 851–4.

Rubinstein, M., Bermúdez, J.-L., Rakov, V. A., Rachidi, F. and Hussein, A. (2012). Compensation of the Instrumental Decay in Measured Lightning Electric Field Waveforms. *IEEE Trans. on EMC*, 54(3) June 2012: 685–8.

Rubinstein, M., Rachidi, F., Uman, M. A., Thottappillil, R., Rakov, V. A. and Nucci, C. A. (1995). Characterization of vertical electric fields 500 m and 30 m from triggered lightning. *J. Geophys. Res.*, 100: 8863–72.

Rubinstein, M. and Uman, M. A. (1989). Methods for calculating the electromagnetic fields from a known source distribution: application to lightning. *IEEE Trans. Electromagn. Compat.*, 31: 183–9.

Rühling F. (1972). Modelluntersuchungen über den Schutzraum und ihre Redeutung für Gebäudeblitzableiter. *Bull. Schweiz. Elektrotech.*, 63: 522–8.

Sadighi, S., Liu, N., Dwyer, J. R. and Rassoul, H. K. (2015). Streamer formation and branching from model hydrometeors in subbreakdown conditions inside thunderclouds. *J. Geophys. Res. Atmos.*, 120, doi:10.1002/2014JD022724.

Sadiku, M. N. O. (1994). *Elements of Electromagnetics*, 821 p., Orlando, FL: Saunders College.

Said, R. K., Cohen, M. B. and Inan, U. S. (2013). Highly intense lightning over the oceans: estimated peak currents from global GLD360 observations. *J. Geophys. Res. Atmos.*, 118: 6905–15, doi:10.1002/jgrd.50508.

Said, R. K., Inan, U. S. and Cummins, K. L. (2010). Long-range lightning geolocation using a VLF radio atmospheric waveform bank. *J. Geophys. Res.*, 115: D23108, doi:10.1029/2010JD013863.

Saraiva, A. C. V., Saba, M. M. F., Pinto, O. Jr., Cummins, K. L., Krider, E. P. and Campos, L. Z. S. (2010). A comparative study of negative cloud-to-ground lightning characteristics in São Paulo (Brazil) and Arizona (United States) based on high-speed video observations. *J. Geophys. Res.*, 115: D11102, doi:10.1029/2009JD012604.

Schoene, J., Uman, M. A., Rakov, V. A., Kodali, V., Rambo, K. J. and Schnetzer, G. H. (2003a). Statistical characteristics of the electric and magnetic fields and their time derivatives 15 m and 30 m from triggered lightning. *J. Geophys, Res.*, 108(D6): 4192, doi:10.1029/2002JD002698.

Schoene, J., Uman, M. A., Rakov, V. A., Rambo, K. J., Jerauld, J. and Schnetzer, G. H. (2003b). Test of the transmission line model and the traveling current source model with triggered lightning return strokes at very close range. *J. Geophys. Res.*, 108(D23): 4737, doi:10.1029/2003JD003683.

Schonland, B. F. J. (1938). Progressive lightning, Pt. 4, The discharge mechanisms. *Proc. Roy. Soc. (London)*, A164: 132–50.

Schoene, J., Uman, M. A., Rakov, V. A., Rambo, K. J., Jerauld, J., Mata, C. T., Mata, A. G., Jordan, D. M. and Schnetzer, G. H. (2009). Characterization of return-stroke currents in rocket-triggered lightning. *J. Geophys. Res.*, 114: D03106, doi:10.1029/2008JD009873.

Schonland, B. F. J., Hodges, D. B. and Collens, H. (1938a). Progressive lightning, pt. 5, A comparison of photographic and electrical studies of the discharge process. *Proc. Roy. Soc. (London)*, A166: 56–75.

Schonland, B. F. J., Malan, D. J. and Collens, H. (1938b). Progressive lightning, pt. 6. *Proc. Roy. Soc. (London)*, A168: 455–69.

Shao, X. M. (1993). The development and structure of lightning discharges observed by VHF radio interferometer. Ph.D. Dissertation, Socorro, New Mexico: New Mexico Inst. of Mining and Technol..

Shao, X. M., Krehbiel, P. R., Thomas, R. J. and Rison, W. (1995). Radio interferometric observations of cloud-to-ground lightning phenomena in Florida. *J. Geophys. Res.*, 100: 2749–83.

Smith, D. A., Shao, X. M., Holden, D. N., Rhodes, C. T., Brook, M., Krehbiel, P. R., Stanley, M., Rison, W. and Thomas, R. J. (1999). A distinct class of isolated intracloud discharges and their associated radio emissions. *J. Geophys. Res.*, 104: 4189–212.

Solomon, R., Schroeder, V. and Baker, M. B. (2001). Lightning initiation – conventional and runaway-breakdown hypotheses. *Q. J. R. Meteorol. Soc.*, 127: 2683–704.

Stock, M. G., Akita, M., Krehbiel, P. R., Rison, W., Edens, H. E., Kawasaki, Z. and Stanley, M. A. (2014). Continuous broadband digital interferometry of lightning using a generalized cross-correlation algorithm. *J. Geophys. Res. Atmos.*, 119: 3134–65, doi:10.1002/2013JD020217.

Stolzenburg, M., Marshall, T. C., Karunarathne, S., Karunarathna, N., Warner, T. A., Orville, R. E. and Betz, H.-D. (2012). Strokes of upward illumination occurring within a few milliseconds after typical lightning return strokes. *J. Geophys. Res.*, 117: D15203, doi:10.1029/2012JD017654.

Stolzenburg, M., Rust, W. D. and Marshall, T. C. (1998a). Electrical structure in thunderstorm convective regions. 2. Isolated storms. *J. Geophys. Res.*, 103: 14, 79–96.

Stolzenburg, M., Rust, W. D. and Marshall, T. C. (1998b). Electrical structure in thunderstorm convective regions. 3. Synthesis. *J. Geophys. Res.*, 103: 14,097–108.

Stolzenburg, M., Rust, W. D., Smull, B. F. and Marshall, T. C. (1998c). Electrical structure in thunderstorm convective regions. 1. Mesoscale convective systems. *J. Geophys. Res.*, 103: 14,059–78.

Szczerbinski, M. (2000). A discussion of "Faraday cage" lightning protection and application to real building structures. *J. Electrostat.*, 48: 145–54.

Takami, J. and Okabe, S. (2007). Observational results of lightning current on transmission towers. *IEEE Trans. Power Del.*, 22: 547–56.

Taylor, W. L. (1978). A VHF technique for space–time mapping of lightning discharge processes. *J. Geophys. Res.*, 83: 3575–83.

Thomas, R. J., Krehbiel, P. R., Rison, W., Hunyady, S. J., Winn, W. P., Hamlin, T. and Harlin, J. (2004). Accuracy of the lightning mapping array. *J. Geophys. Res.*, 109: D14207, doi:10.1029/2004JD004549.

Thomson, E. M. (1999). Exact expressions for electric and magnetic fields from a propagating lightning channel with arbitrary orientation. *J. Geophys. Res.*, 104: 22,293–300.

Thottappillil, R. and Rakov, V. A. (2001). On different approaches to calculating lightning electric fields. *J. Geophys. Res.*, 106: 14,191–205.

Thottappillil, R., Rakov, V. A. and Uman, M. A. (1997). Distribution of charge along the lightning channel: Relation to remote electric and magnetic fields and to return stroke models. *J. Geophys. Res.*, 102: 6887–7006.

Thottappillil, R., Rakov, V. A., Uman, M. A., Beasley, W. H., Master, M. J. and Shelukhin, D. V. (1992). Lightning subsequent-stroke electric field peak greater than the first stroke peak and multiple ground terminations. *J. Geophys. Res.*, 97: 7503–9.

Thottappillil, R., Schoene, J. and Uman, M. A. (2001). Return stroke transmission line model for stroke speed near and equal that of light. *Geophys. Res. Lett.*, 28: 3593–6.

Thottappillil, R. and Uman, M. A. (1993). Comparison of lightning return-stroke models. *J. Geophys. Res.*, 98: 22,903–14.

Tran, M. D. and Rakov, V. A. (2015). When does the lightning attachment process actually begin? *J. Geophys. Res. Atmos.*, 120, 6922–6936, doi:10.1002/2015JD023155.

Tran, M. D., Rakov, V. A., Mallick, S., Dwyer, J. R., Nag, A. and Heckman, S. (2015). A terrestrial gamma-ray flash recorded at the Lightning Observatory in Gainesville, Florida. *J. Atmos. Solar-Terrestr. Phys.*, 136, 86–93, doi:10.1016/j.jastp.2015.10.010.

UL 96A, Underwriters Laboratories (1998). Standard for Installation Requirements for Lightning Protection Systems.

Uman, M. A. (1974). The Earth and its atmosphere as a leaky spherical capacitor. *Am. J. Phys.*, 42: 10335.

Uman, M. A. (1987). *The Lightning Discharge*, 377 p., Orlando, FL: Academic Press.

Uman, M. A. (2001). *The Lightning Discharge*, 377 p., Mineola, NY: Dover.

Uman, M. A. and McLain, D. K. (1969). Magnetic field of the lightning return stroke. *J. Geophys. Res.*, 74: 6899–910.

Uman, M. A., McLain, D. K. and Krider, E. P. (1975). The electromagnetic radiation from a finite antenna. *Am. J. Phys.*, 43: 33–8.

Uman, M. A. and Rakov, V. A. (2002) A critical review of non-conventional approaches to lightning protection, *Bull. Amer. Meteorol. Soc.*, December 2002, 1809–20.

Valine, W. C. and Krider, E. P. (2002). Statistics and characteristics of cloud-to-ground lightning with multiple ground contacts. *J. Geophys. Res.*, 107(D20): 4441, doi:10.1029/2001JD001360.

Vance, E. F. (1980). Electromagnetic interference control. *IEEE Trans. Electromagn. Compat.*, 22: 54–62.

Visacro, S., Mesquita, C. R., De Conti, A. and Silveira, F. H. (2012). Updated statistics of lightning currents measured at Morro do Cachimbo station. *Atmos. Res.*, 117: 55–63.

Wagner, C. F. (1963). Relation between stroke current and velocity of the return stroke. *AIEE Trans Pow. Appar. Syst.*, 82: 609–17.

Walsh, K., Cooper, M. A., Holle, R., Rakov, V. A., Roeder, W. P. and Ryan, M. (2013). National Athletic Trainers' Association position statement: Lightning safety for athletics and recreation. *Journal of Athletic Training*, 48(2): 258–70, doi:10.4085/1062-6050-48.2.25.

Wang, D., Rakov, V. A., Uman, M. A., Takagi, N., Watanabe, T., Crawford, D. E., Rambo, K. J., Schnetzer, G. H., Fisher, R. J. and Kawasaki, Z.-I. (1999a). Attachment process in rocket-triggered lightning strokes. *J Geophys Res.*, 104: 2143–50.

Wang, D., Takagi, N., Watanabe, T., Rakov, V. A. and Uman, M. A. (1999b). Observed leader and return-stroke propagation characteristics in the bottom 400 m of a rocket-triggered lightning channel. *J. Geophys. Res.*, 104: 14369–76.

Wang, Y., Zhang, G., Qie, X., Wang, D., Zhang, T., Zhao, Y., Li, Y. and Zhang, T. (2012). Characteristics of compact intracloud discharges observed in a severe thunderstorm in northern part of China. *Journal of Atmospheric and Solar-Terrestrial Physics*, 84–5, 7–14.

Weidman, C. D. and Krider, E. P. (1978). The fine structure of lightning return stroke wave forms. *J. Geophys. Res.*, 83: 6239–47. (Correction 1982, 87: 7351).

Weidman, C. D. and Krider, E. P. (1979). The radiation field wave forms produced by intracloud lightning discharge processes. *J. Geophys. Res.*, 84: 3159–64.

Wiesinger, J. and Zischank, W. (1995). Lightning protection. In Volland, H. (ed.), *Handbook of Atmospheric Electrodynamics*, vol. II, 33–64. Boca Raton, FL: CRC Press.

Willett, J. C., Bailey, J. C., Idone, V. P., Eybert-Berard, A. and Barret, L. (1989). Submicrosecond intercomparison of radiation fields and currents in triggered lightning return strokes based on the transmission line model. *J. Geophys. Res.*, 94: 13, 27586.

Williams, E. R. (1995). Meteorological aspects of thunderstorms. In Volland, H. (ed.), *Handbook of Atmospheric Electrodynamics*, vol. I, 27–60. Boca Raton, FL: CRC Press.

Wilson, N., Myers, J., Cummins, K., Hutchinson, M. and Nag, A. (2013). Lightning attachment to wind turbines in Central Kansas: Video observations, correlation with the NLDN and in situ peak current measurements. *European Wind Energy Association (EWEA)*, Vienna. Austria, 4–7 February, 8.

Wilson, C. T. R. (1920). Investigations on lightning discharges and on the electric field of thunderstorms. *Phil. Trans. Roy. Soc. (London)*, A221: 73–115.

WMO Publication 21. (1956). World Distribution of Thunderstorm Days, Part 2, Tables of Marine Data and World Maps. Geneva, Switzerland.

Wu, T., Dong, W., Zhang, Y., Funaki, T., Yoshida, S., Morimoto, T., Ushio, T. and Kawasaki, Z. (2012). Discharge height of lightning narrow bipolar events. *J. Geophys. Res.*, 117: D05119, doi:10.1029/2011JD017054.

Wu, T., Takayanagi, Y., Yoshida, S., Funaki, T., Ushio, T. and Kawasaki, Z. (2013). Spatial relationship between lightning narrow bipolar events and parent thunderstorms as revealed by phased array radar. *Geophys. Res. Lett.*, 40, doi:10.1002/grl.50112.

Wu, T., Yoshida, S., Ushio, T., Kawasaki, Z. and Wang, D. (2014). Lightning-initiator type of narrow bipolar events and their subsequent pulse trains. *J. Geophys. Res. Atmos.*, 119(12): 7425–38, doi:10.1002/2014JD021842.

Zhu, Y., Rakov, V. A., Tran, M. D. and Nag, A. (2016). A study of NLDN responses to cloud discharge activity based on ground-truth data acquired at the LOG, 2016 ILDC/ILMC, San Diego, California, USA.

Index

Printed in the United States
by Baker & Taylor Publisher Services